Healthcare Compliance Officer Study Guide

Complete Certification Exam Prep with 300 Practice Questions, Detailed Answer Explanations, and Privacy, Fraud Prevention, Risk Management Essentials

Ochola Sabrina McKnight

First Edition: 2025
ISBN: 978-1-7642339-4-1

CHC, CCEP, HIPAA, Medicare, Medicaid, and other product names mentioned are trademarks of their respective organizations. This book is independent and not affiliated with any certification body.

This book is for educational and exam preparation purposes only. It does not constitute legal, medical, or professional advice. Healthcare compliance involves complex regulations that vary by jurisdiction and circumstance.

- No guarantee of exam success or certification
- No warranty of accuracy, completeness, or currency of information
- Healthcare regulations change frequently - verify current requirements
- Consult qualified legal counsel and current official sources for specific guidance

All personal names, case studies, and organizational examples used in this book are fictitious. Any resemblance to actual persons, organizations, or events is purely coincidental.

Healthcare regulations change through new legislation, agency guidance, court decisions, and enforcement actions. Users must stay current with applicable regulatory changes.

This study guide is independent and not endorsed by HCCA, SCCE, AHIMA, or any certification organization.

Table of Contents

Chapter 1: Introduction to Healthcare Compliance

Healthcare compliance stands as the cornerstone of modern medical practice, protecting patients while safeguarding organizations from regulatory pitfalls. This discipline has transformed from a reactive approach to fraud prevention into a proactive strategy that drives quality care and operational excellence. Understanding this evolution provides the foundation for anyone considering a career in healthcare compliance or seeking to strengthen their organization's regulatory framework.

The Evolution of Healthcare Compliance

The journey of healthcare compliance began in earnest during the 1990s when Medicare fraud initiatives exposed widespread billing irregularities across the healthcare system. Before this period, healthcare organizations operated with minimal oversight, relying primarily on professional ethics and basic billing procedures. The discovery of systematic fraud patterns changed everything.

The Healthcare Insurance Portability and Accountability Act (HIPAA) of 1996 marked the first major legislative milestone. This law introduced privacy protections for patient information and established penalties for violations. Following HIPAA, the Balanced Budget Act of 1997 created the Medicare Integrity Program, giving the Centers for Medicare & Medicaid Services (CMS) enhanced authority to combat fraud and abuse.

The Sarbanes-Oxley Act of 2002, while primarily focused on corporate financial reporting, extended its reach into healthcare by requiring public healthcare companies to implement stronger internal controls. This legislation introduced the concept of executive accountability for compliance programs, making leadership personally responsible for organizational integrity.

The Affordable Care Act of 2010 brought another wave of compliance requirements, introducing value-based care models and quality reporting mandates. This legislation shifted the focus from volume-based to value-based care, creating new compliance challenges around quality metrics and population health management.

Case Example 1: The Tenet Healthcare Settlement In 2006, Tenet Healthcare Corporation paid $900 million to settle allegations of Medicare fraud. The case involved unnecessary cardiac procedures, inflated bills, and kickback schemes. The settlement required Tenet to implement a Corporate Integrity Agreement (CIA) with comprehensive compliance monitoring. This case demonstrated how compliance failures could result in massive financial penalties and long-term government oversight.

The Patient Protection and Affordable Care Act expanded compliance requirements further by introducing the Physician Payments Sunshine Act, which mandates transparency in financial relationships between healthcare providers and pharmaceutical companies. This requirement created new documentation and reporting obligations for healthcare organizations.

Modern compliance programs in 2025 address multiple regulatory frameworks simultaneously. Organizations must navigate HIPAA privacy and security requirements, fraud and abuse laws, quality reporting mandates, and emerging areas like information blocking and telehealth regulations. The complexity has grown exponentially, requiring dedicated professionals with specialized knowledge.

Technology has transformed compliance monitoring from manual chart reviews to sophisticated data analytics platforms. Artificial intelligence now helps identify billing patterns that might indicate fraud, while automated systems track training completion and policy acknowledgments. These technological advances have made compliance more efficient but also more complex to manage.

Why Healthcare Compliance Matters

Patient protection forms the primary purpose of healthcare compliance. Every regulation, policy, and procedure exists to ensure patients receive safe, appropriate, and ethical care. Compliance programs protect vulnerable populations from exploitation while maintaining the integrity of the healthcare system.

The financial implications of compliance failures extend far beyond immediate penalties. Organizations face direct costs including fines, settlements, and legal fees. Indirect costs include lost revenue during investigations, increased insurance premiums, and reduced referrals due to reputation damage. A single compliance failure can cost millions of dollars and take years to recover from fully.

Case Example 2: The Anthem Data Breach In 2015, Anthem Inc. experienced a cyberattack that compromised 78.8 million patient records. The company paid $16 million to settle HIPAA violations with the Office for Civil Rights (OCR). Beyond the direct penalty, Anthem spent over $260 million on remediation efforts, legal costs, and credit monitoring services for affected individuals. The breach highlighted how cybersecurity failures create cascading compliance violations with massive financial consequences.

Legal and regulatory requirements create a complex web of obligations that healthcare organizations must navigate carefully. Federal agencies including CMS, the Office of Inspector General (OIG), and the Food and Drug Administration (FDA) each have enforcement authority over different aspects of healthcare operations. State agencies add another layer of requirements, creating challenges for multi-state organizations.

Organizational reputation represents one of the most valuable assets healthcare organizations possess. Compliance failures generate negative media coverage, erode public trust, and damage relationships with referring physicians. Patients often choose healthcare providers based on perceived quality and safety, making reputation management a critical business concern.

The sustainability of healthcare organizations depends on their ability to maintain compliance over time. Organizations with strong compliance programs demonstrate their commitment to ethical practices, making them more attractive to investors, partners, and employees. Compliance excellence creates competitive advantages in an increasingly regulated industry.

Case Example 3: The University of Miami Health System Success Story The University of Miami Health System transformed its compliance program following a 2011 investigation. The organization implemented comprehensive policies, enhanced training programs, and created a culture of transparency. By 2020, the system had achieved recognition as a compliance leader, attracting top talent and improving patient satisfaction scores. This transformation demonstrates how effective compliance programs create long-term organizational value.

Quality of care improvements often result from effective compliance programs. When organizations focus on appropriate utilization, proper documentation, and evidence-based practices, patient outcomes naturally improve. Compliance creates accountability mechanisms that drive clinical excellence throughout the organization.

The Role of Healthcare Compliance Officers

Healthcare compliance officers serve as the guardians of organizational integrity, balancing regulatory requirements with operational efficiency. These professionals navigate complex regulatory environments while building cultures of ethical practice throughout their organizations.

Core responsibilities include developing and maintaining compliance policies, conducting risk assessments, and overseeing audit activities. Compliance officers design training programs that educate staff about regulatory requirements and ethical decision-making. They investigate potential violations, coordinate corrective actions, and report findings to senior leadership and government agencies when required.

Daily activities vary significantly based on organizational size and complexity. Large health systems employ teams of compliance professionals with specialized roles, while smaller organizations may have a single compliance officer handling all functions. Typical daily tasks include reviewing incident reports, updating policies based on regulatory changes, conducting training sessions, and meeting with department leaders about compliance issues.

The compliance officer's position within organizational structure affects their effectiveness significantly. Best practices recommend reporting directly to the CEO or board of directors to ensure independence and authority. This reporting relationship allows compliance officers to address issues without fear of retaliation and provides access to resources needed for effective program implementation.

Communication skills are essential for compliance officers, who must translate complex regulations into practical guidance for clinical and administrative staff. They serve as internal consultants, helping departments understand how regulations apply to their specific functions. Effective compliance officers build collaborative relationships across organizations rather than serving as compliance police.

Required competencies include regulatory knowledge, analytical thinking, and project management skills. Healthcare compliance officers must understand clinical operations, financial management, and information technology systems. They need knowledge of healthcare law, audit techniques, and risk management principles. Strong written and verbal communication skills are essential for policy development, training delivery, and investigation reporting.

Career paths in healthcare compliance offer diverse opportunities for professional growth. Entry-level positions include compliance analysts, audit coordinators, and privacy officers. Mid-level roles encompass compliance managers, HIPAA security officers, and research compliance coordinators. Senior positions include chief compliance officers, vice presidents of compliance, and consultant roles.

Geographic considerations affect career opportunities and compensation levels. Major metropolitan areas typically offer more positions and higher salaries but also increased competition. Rural areas may have fewer opportunities but offer broader responsibilities and faster career advancement. Remote work options have expanded significantly, particularly for consultant and audit roles.

Overview of Certification Options

Professional certifications validate expertise and demonstrate commitment to healthcare compliance excellence. Multiple organizations offer certifications tailored to different aspects of healthcare compliance, each with specific requirements and benefits.

The Certified in Healthcare Compliance (CHC) credential from the Health Care Compliance Association (HCCA) provides the most extensive coverage of healthcare compliance topics. This certification addresses fraud and abuse laws, HIPAA requirements, quality reporting, and compliance program management. CHC certification requires relevant work experience, continuing education, and successful completion of a challenging examination.

The Certified Compliance and Ethics Professional - Healthcare (CCEP-HC) from the Society of Corporate Compliance and Ethics focuses on ethics and compliance program development. This certification emphasizes organizational culture, risk assessment, and program effectiveness measurement. CCEP-HC candidates must demonstrate leadership experience and commitment to ethical practices.

The Certified in Healthcare Research Compliance (CHRC) addresses the specialized field of research compliance. This certification covers human subjects protection, good clinical practice guidelines, and regulatory requirements for clinical trials. CHRC certification is ideal for professionals working in academic medical centers, pharmaceutical companies, or contract research organizations.

The Certified in Healthcare Privacy Compliance (CHPC) focuses specifically on HIPAA privacy requirements and related state laws. This certification addresses privacy risk assessment, breach notification, and business associate management. CHPC certification suits privacy officers and professionals with specialized privacy responsibilities.

The Certified in Healthcare Privacy and Security (CHPS) from the American Health Information Management Association (AHIMA) combines privacy and security competencies. This certification covers both HIPAA Privacy and Security Rules, including technical safeguards and risk analysis requirements. CHPS certification is particularly valuable for health information management professionals.

Certification comparison requires careful consideration of career goals, current responsibilities, and long-term objectives. The CHC offers the broadest foundation for healthcare compliance careers, while specialized certifications like CHRC or CHPS provide deeper expertise in specific areas. Many professionals pursue multiple certifications throughout their careers to demonstrate expertise in various compliance domains.

Preparation strategies vary by certification but generally include formal education, practical experience, and focused study. Most certifications require continuing education to maintain credentials, ensuring professionals stay current with regulatory changes and best practices. Certification maintenance demonstrates ongoing commitment to professional development and regulatory expertise.

Chapter Exercises

Self-Assessment: Is Healthcare Compliance Right for You?

Healthcare compliance careers require specific personality traits and professional interests. Effective compliance professionals enjoy solving complex problems, working with detailed regulations, and helping organizations improve their practices. They must be comfortable with ambiguity, as regulations often require interpretation and judgment.

Consider your tolerance for detailed work, as compliance involves reviewing policies, analyzing data, and documenting findings. Evaluate your communication skills, as compliance officers must explain complex concepts to diverse audiences. Assess your ethical standards, as compliance work requires absolute integrity and independence.

Think about your interest in healthcare operations and regulatory requirements. Successful compliance professionals stay current with changing regulations and understand how they affect clinical and business operations. They find satisfaction in preventing problems rather than just responding to them.

Certification Pathway Planning Worksheet

Begin by assessing your current education, experience, and career goals. Different certifications have varying eligibility requirements, so understanding your background helps identify appropriate options. Consider your current role and how certification might enhance your responsibilities or advance your career.

Research certification requirements including education, experience, and examination components. Some certifications require specific types of healthcare experience, while others accept related compliance work. Understanding these requirements helps you plan your professional development activities.

Create a timeline for certification pursuit, including preparation time, examination scheduling, and continuing education requirements. Most professionals need several months to prepare adequately for certification examinations. Factor in your current workload and personal commitments when planning your certification timeline.

Key Terms Glossary Builder

Healthcare compliance involves extensive terminology that professionals must master. Begin building your glossary with fundamental terms like compliance program, risk assessment, and corrective action. Add regulatory terms such as designated health services, protected health information, and corporate integrity agreement.

Include enforcement terms like qui tam lawsuit, exclusion, and civil monetary penalty. Add quality-related terms such as quality measure, value-based care, and patient safety indicator. Understanding this terminology provides the foundation for effective communication in healthcare compliance.

Regularly update your glossary as you encounter new terms and concepts. Healthcare compliance terminology evolves with regulatory changes and industry developments. Maintaining a current glossary serves as a valuable reference throughout your compliance career.

Healthcare compliance has evolved from a necessary burden into a strategic advantage for forward-thinking organizations. The professionals who master this discipline become trusted advisors, helping their organizations navigate an increasingly complex regulatory environment while maintaining focus on patient care and organizational mission.

The path forward requires dedication to continuous learning, ethical practice, and collaborative problem-solving. Each regulation mastered, each policy developed, and each training session delivered contributes to a healthcare system that better serves patients and communities. This is work that matters—not just for organizational success, but for the fundamental trust that patients place in healthcare providers every day.

As you consider your role in this essential field, think about the impact you want to make. Healthcare compliance professionals protect vulnerable patients, preserve organizational integrity, and contribute to a healthcare system worthy of public trust. These are responsibilities that require both technical expertise and moral courage, offering rewards that extend far beyond professional achievement.

Key Takeaways

- Healthcare compliance evolved from 1990s fraud initiatives into a proactive discipline that drives quality care and organizational excellence
- Compliance failures create cascading financial, legal, and reputational consequences that can threaten organizational sustainability
- Healthcare compliance officers serve as organizational guardians, requiring diverse skills including regulatory knowledge, communication abilities, and ethical leadership
- Multiple certification options exist, with CHC providing the broadest foundation and specialized credentials offering focused expertise
- Success in healthcare compliance requires continuous learning, ethical practice, and commitment to patient protection above all other considerations

Chapter 2: Healthcare Compliance Program Infrastructure

Building an effective healthcare compliance program requires more than good intentions and regulatory knowledge. Organizations need structured frameworks, clear accountability mechanisms, and sustainable processes that integrate seamlessly with daily operations. The infrastructure that supports compliance activities determines whether programs succeed in preventing violations or merely create paperwork burdens that add cost without value.

The Seven Elements of an Effective Compliance Program

The foundation of healthcare compliance rests on seven essential elements established by the Office of Inspector General (OIG) in its compliance program guidance. These elements work together to create a systematic approach to preventing, detecting, and correcting compliance violations.

Written Policies, Procedures, and Standards of Conduct

Written policies serve as the roadmap for ethical decision-making throughout healthcare organizations. These documents translate complex regulations into practical guidance that employees can understand and follow. Effective policies address specific situations that employees encounter while providing enough flexibility to handle unique circumstances.

Standards of conduct establish the ethical framework for organizational behavior. These documents communicate expectations for all employees, from entry-level staff to senior executives. They address conflicts of interest, reporting obligations, and consequences for violations. The most effective standards of conduct use clear language and real-world examples rather than legal jargon.

Policy development requires careful balance between regulatory compliance and operational efficiency. Policies that are too restrictive can interfere with patient care, while those that are too permissive may not prevent violations. Regular review and update processes ensure policies remain current with regulatory changes and organizational needs.

Case Example 1: The Mayo Clinic Policy Framework Mayo Clinic has developed one of healthcare's most comprehensive policy frameworks, with over 2,000 policies covering clinical, administrative, and compliance topics. Their system includes automated alerts when regulations change, requiring policy owners to review and update relevant documents within 90 days. This proactive approach has helped Mayo Clinic maintain an exemplary compliance record while supporting clinical innovation.

Compliance Officer, Compliance Committee, and High-Level Oversight

The compliance officer serves as the program's operational leader, responsible for day-to-day implementation and oversight activities. This position requires independence from operational pressures, typically achieved through direct reporting to the CEO or board of directors. Compliance officers need sufficient authority to investigate issues, access information, and implement corrective actions.

Compliance committees provide governance oversight and strategic direction for compliance programs. These committees typically include senior executives from clinical, financial, legal, and operational departments. Committee composition should reflect organizational structure and risk profile, ensuring all major areas have representation and input.

Board oversight establishes tone at the top and demonstrates organizational commitment to compliance. Board members need regular compliance reporting, including metrics on program effectiveness, investigation results, and corrective actions. Effective boards ask probing questions and hold leadership accountable for compliance performance.

High-level oversight creates accountability throughout the organization. This includes regular reporting to senior leadership, board presentations on compliance activities, and integration of compliance metrics into performance evaluations. Organizations that treat compliance as a senior leadership priority achieve better results than those that delegate it to lower levels.

Effective Training and Education

Training programs ensure employees understand their compliance obligations and know how to fulfill them. Effective training goes beyond annual presentations to include role-specific education, case-based learning, and ongoing reinforcement. Training should address both general compliance principles and specific regulatory requirements relevant to each employee's responsibilities.

New employee orientation provides the first opportunity to establish compliance expectations. This training should cover organizational values, standards of conduct, reporting mechanisms, and consequences for violations. Role-specific training during orientation ensures employees understand compliance requirements for their particular positions.

Ongoing education keeps employees current with regulatory changes and reinforces compliance messages. This includes regular updates on new regulations, lessons learned from investigations, and emerging risk areas. The most effective programs use multiple delivery methods including online modules, in-person sessions, and informal communications.

Training effectiveness requires measurement and evaluation. Organizations should track completion rates, test scores, and behavior changes following training interventions. Regular surveys can assess employee understanding and identify areas needing additional emphasis.

Case Example 2: The Cleveland Clinic Training Innovation
Cleveland Clinic developed a gamified compliance training platform that increased completion rates from 65% to 95% within one year. The system uses interactive scenarios, point-based rewards, and leaderboards to engage employees. More importantly, the organization documented improved compliance behavior and reduced violation rates following implementation of the new training approach.

Effective Lines of Communication

Communication systems enable employees to seek guidance, report concerns, and receive feedback about compliance matters. These systems must be accessible, confidential, and responsive to encourage utilization. Multiple communication channels accommodate different employee preferences and situations.

Hotlines provide anonymous reporting mechanisms for employees who observe potential compliance violations. Effective hotlines are available 24/7, staffed by trained personnel, and managed by independent third parties to ensure confidentiality. Organizations should regularly promote hotline availability and demonstrate that reports receive thorough investigation.

Open-door policies encourage direct communication between employees and compliance staff. These policies work best when compliance officers are visible throughout the organization and approachable for informal discussions. Regular office hours, department visits, and informal meetings help build relationships that facilitate communication.

Feedback mechanisms close the communication loop by informing employees about investigation results and corrective actions. While maintaining appropriate confidentiality, organizations should communicate lessons learned and system improvements resulting from employee reports. This demonstrates that reports are taken seriously and lead to meaningful changes.

Internal Monitoring and Auditing

Monitoring and auditing activities detect compliance violations before they become serious problems. These activities should be risk-based, focusing on areas with highest probability and impact of violations. Regular monitoring creates deterrent effects while providing early warning of emerging issues.

Audit programs should include both routine reviews and targeted investigations based on risk indicators. Routine audits cover high-risk areas on regular schedules, while targeted audits respond to specific concerns or trending issues. Audit scope should encompass clinical, financial, and administrative activities across all organizational locations.

Data analytics enhance monitoring effectiveness by identifying patterns and anomalies that might indicate compliance problems. Modern analytics platforms can analyze billing patterns, prescribing behavior, and documentation quality to flag potential issues. These tools allow organizations to focus audit resources on the highest-risk areas.

Audit findings must be documented, communicated, and addressed through corrective actions. Audit reports should clearly describe violations found, root causes identified, and recommendations for improvement. Leadership should receive regular summaries of audit activities and findings to maintain oversight of compliance performance.

Response to Detected Offenses and Corrective Action

When compliance violations occur, organizations must respond promptly and effectively to address the immediate problem and prevent recurrence. Response protocols should be established in advance to ensure consistent, appropriate actions regardless of who is involved or what type of violation occurred.

Investigation procedures guide the process of gathering facts, interviewing witnesses, and documenting findings. These procedures should ensure thoroughness while protecting employee rights and maintaining confidentiality. Legal counsel should be involved in investigations that might result in regulatory reporting or legal liability.

Corrective actions address both the specific violation and underlying causes that allowed it to occur. These actions might include policy updates, training enhancements, system changes, or disciplinary measures. The most effective corrective actions prevent similar violations rather than just addressing the immediate problem.

Self-disclosure to government agencies may be required or advisable depending on violation severity and organizational circumstances. Organizations should have established protocols for evaluating disclosure obligations and managing government interactions. Early, voluntary disclosure often results in reduced penalties and demonstrates good faith compliance efforts.

Case Example 3: The Banner Health Response Model When Banner Health discovered billing irregularities in 2018, they implemented a comprehensive response that included immediate corrective actions, voluntary disclosure to government agencies, and systematic review of related areas. Their transparent approach and thorough corrective action plan resulted in a favorable settlement agreement and recognition for cooperation. The organization used the experience to strengthen their compliance program and create a template for future incident response.

Enforcement of Disciplinary Standards

Disciplinary standards ensure that compliance violations have meaningful consequences, creating accountability for all employees regardless of position or performance. These standards should be clearly communicated, consistently applied, and proportionate to violation severity.

Progressive discipline policies outline escalating consequences for repeated or serious violations. These policies typically include verbal warnings, written warnings, suspension, and termination as potential responses. The specific response should consider violation severity, employee intent, cooperation with investigations, and prior compliance history.

Leadership accountability requires that senior employees face appropriate consequences for compliance failures within their areas of responsibility. This includes both direct violations and failures to prevent or detect violations by subordinates. Holding leaders accountable demonstrates organizational commitment to compliance at all levels.

Documentation of disciplinary actions protects organizations and ensures consistency over time. This documentation should include facts supporting the discipline, rationale for the specific action taken, and any mitigating or aggravating factors considered. Proper documentation also supports defense against wrongful termination claims or discrimination allegations.

Organizational Structure for Compliance

The structure of compliance programs affects their effectiveness and sustainability. Well-designed structures provide clear authority, appropriate resources, and effective oversight while integrating seamlessly with existing organizational systems.

Board Oversight Responsibilities

Board oversight establishes compliance as an organizational priority and provides independent review of program effectiveness. Boards should receive regular reports on compliance activities, investigation results, and program metrics. They should also approve major policy changes and resource allocations for compliance activities.

Board education ensures directors understand their compliance oversight responsibilities and the regulatory environment affecting their organization. This includes training on healthcare fraud and abuse laws, quality reporting requirements, and governance best practices. Well-informed boards provide more effective oversight and ask better questions about compliance performance.

Compliance committee structures at the board level vary based on organizational size and complexity. Some organizations have dedicated compliance committees, while others integrate compliance oversight into audit or quality committees. The key is ensuring adequate time and attention for compliance matters at the highest governance level.

Compliance Committee Composition and Functions

Organizational compliance committees provide operational oversight and strategic direction for compliance programs. These committees should include representatives from all major organizational functions including clinical, financial, legal, information technology, and human resources departments.

Committee functions include policy review and approval, resource allocation decisions, investigation oversight, and corrective action monitoring. Committees should meet regularly, maintain documented minutes, and report to senior leadership and the board about their activities and decisions.

Committee effectiveness depends on member engagement, meeting preparation, and follow-through on decisions. Members should come prepared to discuss compliance issues within their areas of responsibility and commit resources to address identified problems. Committee decisions should be documented and communicated throughout the organization.

Reporting Relationships and Independence

Compliance officer reporting relationships significantly affect program independence and effectiveness. Direct reporting to the CEO or board provides maximum independence but may create communication challenges with operational departments. Reporting through other executives can create conflicts when compliance requirements conflict with operational goals.

Dotted-line reporting relationships can provide operational coordination while maintaining compliance independence. In these structures, compliance officers report formally to the CEO or board but have regular communication and coordination responsibilities with other senior executives.

Functional independence allows compliance officers to investigate issues and recommend corrective actions without fear of retaliation or interference. This independence should be clearly established in organizational policies and reinforced through performance evaluation and compensation structures.

Building a Culture of Compliance

Organizational culture determines whether compliance programs succeed in preventing violations or merely create documentation after problems occur. Strong compliance cultures integrate ethical decision-making into daily operations rather than treating compliance as a separate function.

Tone at the Top

Leadership behavior establishes the foundation for organizational compliance culture. When leaders consistently demonstrate commitment to compliance through their words and actions, employees follow their example. Conversely, leaders who ignore compliance requirements or prioritize financial results over ethical behavior create cultures that tolerate violations.

Leadership communication should consistently reinforce compliance messages and demonstrate personal commitment to ethical behavior. This includes regular communication about compliance expectations, recognition of employees who demonstrate compliance excellence, and visible consequences for violations regardless of employee position.

Executive participation in compliance activities demonstrates leadership commitment and importance. This includes attending training sessions, participating in compliance committee meetings, and personally addressing compliance concerns when they arise. Leaders who are visibly engaged in compliance activities create more effective compliance cultures.

Employee Engagement Strategies

Employee engagement in compliance requires making compliance relevant to daily work activities rather than treating it as additional burden. This includes showing how compliance protects patients, supports organizational mission, and creates better work environments for all employees.

Recognition programs acknowledge employees who demonstrate compliance excellence through their daily activities. These programs should highlight specific behaviors that exemplify compliance values rather than just recognizing perfect attendance at training sessions. Public recognition reinforces positive behaviors and encourages others to follow similar approaches.

Feedback mechanisms allow employees to contribute ideas for compliance improvement and receive responses to their suggestions. Organizations that actively seek employee input and implement good suggestions create more engaged compliance cultures than those that treat compliance as top-down mandate.

Communication Frameworks

Effective communication frameworks ensure compliance messages reach all employees through appropriate channels and formats. Different employee groups may prefer different communication methods, requiring multiple approaches to achieve comprehensive coverage.

Regular communication schedules maintain compliance visibility throughout the year rather than concentrating messages during annual training periods. This includes monthly newsletters, quarterly updates, and special communications about emerging issues or regulatory changes.

Two-way communication opportunities allow employees to ask questions, seek clarification, and provide feedback about compliance matters. Regular office hours, department visits, and informal discussions create opportunities for meaningful dialogue about compliance challenges and solutions.

Documentation and Record-Keeping Requirements

Proper documentation supports compliance program effectiveness while meeting regulatory requirements for record retention and government reporting. Documentation systems should balance thoroughness with efficiency, creating necessary records without imposing excessive administrative burdens.

Policy Management Systems

Policy management systems organize, maintain, and distribute organizational policies while tracking revisions and ensuring currency. These systems should provide version control, automated review reminders, and distribution tracking to ensure employees have access to current policies.

Electronic policy systems offer advantages including search capabilities, automated notifications, and usage tracking. These systems can remind policy owners about scheduled reviews, track employee acknowledgments, and provide analytics about policy utilization patterns.

Policy lifecycle management includes development, review, approval, distribution, and retirement processes. Each stage should have clear responsibilities, timelines, and quality controls to ensure policies remain current and useful for their intended purposes.

Training Documentation

Training documentation demonstrates compliance with regulatory requirements while supporting program evaluation and improvement efforts. This documentation should include training content, attendance records, test results, and evaluation feedback.

Learning management systems automate training documentation while providing analytics about program effectiveness. These systems can track completion rates, test scores, and time to completion while generating reports for compliance officers and government agencies.

Training records retention should comply with regulatory requirements and organizational policies. Different types of training may have different retention requirements, requiring systematic record management to ensure appropriate preservation and disposal.

Investigation Records

Investigation documentation serves multiple purposes including regulatory compliance, legal protection, and organizational learning. These records should thoroughly document investigation processes, findings, and corrective actions while protecting employee privacy and confidentiality.

Investigation files should include initial complaints or reports, evidence gathered, interviews conducted, findings reached, and corrective actions implemented. Documentation should be sufficient to support conclusions reached and demonstrate thorough investigation processes.

Confidentiality protections for investigation records require careful balance between transparency and privacy. Records should be maintained securely with access limited to individuals with legitimate business needs. Legal counsel should guide decisions about record disclosure and retention.

Practical Implementation Strategies

Healthcare compliance programs require practical implementation approaches that integrate regulatory requirements with operational realities. Successful programs adapt theoretical frameworks to organizational culture, resources, and risk profiles.

Phased implementation allows organizations to build compliance programs systematically rather than attempting to implement all elements simultaneously. This approach permits learning and adjustment while avoiding overwhelming employees with too many changes at once.

Resource allocation decisions should prioritize highest-risk areas while ensuring adequate coverage of all compliance domains. Organizations with limited resources should focus initial efforts on areas with greatest regulatory scrutiny and potential penalties.

Change management strategies help employees adapt to new compliance requirements and processes. This includes clear communication about reasons for changes, training on new procedures, and support during transition periods. Effective change management reduces resistance and accelerates compliance program adoption.

Moving Forward Together

Healthcare compliance program infrastructure requires careful planning, adequate resources, and sustained commitment from organizational leadership. The seven elements of effective compliance programs provide a proven framework, but each organization must adapt these elements to their unique circumstances and challenges.

Building compliance infrastructure is an investment in organizational sustainability and patient protection. While the initial effort may seem overwhelming, the systematic approach outlined in these elements creates programs that prevent problems rather than just responding to them after they occur. This proactive approach ultimately saves resources while better serving patients and communities.

The journey toward compliance excellence begins with commitment to the foundational elements described here. Each policy written, each training session delivered, and each investigation conducted contributes to a culture that values ethical behavior and regulatory compliance. This is infrastructure that supports not just compliance, but organizational excellence in all its forms.

Key Takeaways

- The seven elements of effective compliance programs provide a systematic framework that must be tailored to each organization's unique circumstances and risk profile
- Organizational structure significantly affects compliance program effectiveness, with independence and authority being critical success factors
- Building a culture of compliance requires sustained leadership commitment demonstrated through words, actions, and resource allocation decisions

- Documentation and record-keeping systems must balance regulatory compliance with operational efficiency to avoid creating administrative burdens
- Successful implementation requires phased approaches, adequate resources, and effective change management to achieve sustainable compliance excellence

Chapter 3: HIPAA Compliance Mastery

The Health Insurance Portability and Accountability Act stands as one of healthcare's most influential regulations, fundamentally changing how organizations handle patient information. Since its implementation, HIPAA has evolved from basic privacy protections into a complex framework addressing privacy, security, and breach notification requirements. Understanding HIPAA's intricacies is essential for healthcare compliance professionals, as violations can result in massive penalties and lasting reputational damage.

HIPAA Privacy Rule Deep Dive

The Privacy Rule establishes national standards for protecting individually identifiable health information, creating the foundation for patient privacy rights that most Americans now take for granted. This rule affects every healthcare organization, from large hospital systems to individual practitioners, requiring systematic approaches to information protection.

Protected Health Information Identification

Protected Health Information (PHI) encompasses any individually identifiable health information held or transmitted by covered entities and their business associates. Understanding what constitutes PHI requires careful analysis, as the definition extends beyond obvious examples like medical records to include seemingly innocuous information that could identify patients.

The 18 identifiers that make health information individually identifiable include names, addresses, telephone numbers, email addresses, social security numbers, medical record numbers, health plan beneficiary numbers, account numbers, certificate or license numbers, vehicle identifiers, device identifiers, web URLs, IP addresses, biometric identifiers, full-face photographs, and any other unique identifying number or characteristic.

De-identification processes remove these identifiers to create information that no longer qualifies as PHI. The Privacy Rule provides two methods for de-identification: the safe harbor method, which requires removal of specific identifiers, and the expert determination method, which relies on statistical analysis to determine re-identification risk.

Limited data sets represent a middle ground between PHI and de-identified information. These data sets may contain dates, city, state, ZIP codes, and ages over 89, but must be used only for research, public health, or healthcare operations purposes under formal data use agreements.

Case Example 1: The University of Rochester Medical Center Violation In 2019, the University of Rochester Medical Center paid $3 million to settle HIPAA violations involving the disclosure of PHI to a documentary film crew. The organization failed to obtain proper authorization before allowing filming in areas where PHI was visible, including patient rooms and medical records. This case demonstrates how seemingly routine activities can create privacy violations when proper safeguards are not implemented.

Minimum Necessary Standards

The minimum necessary standard requires covered entities to limit PHI uses and disclosures to the minimum amount reasonably necessary to accomplish the intended purpose. This principle applies to routine disclosures, requests for information, and internal access to patient information.

Implementation of minimum necessary standards requires policies and procedures that define what information different workforce members need for their job functions. These policies should consider job responsibilities, patient care needs, and operational requirements while limiting access to unnecessary information.

Role-based access controls in electronic systems help enforce minimum necessary standards by automatically limiting information access based on user roles and responsibilities. These controls should be regularly reviewed and updated as job functions change or new roles are created.

Training programs must address minimum necessary requirements, helping employees understand how to apply these standards in their daily work. This includes guidance on information sharing decisions, request evaluation, and system access appropriate for specific job functions.

Patient Rights and Access

The Privacy Rule grants patients specific rights regarding their health information, including access, amendment, accounting of disclosures, and restriction requests. Healthcare organizations must have processes to honor these rights while managing operational impacts.

Access rights allow patients to inspect and obtain copies of their PHI maintained in designated record sets. Organizations must respond to access requests within 30 days, with one 30-day extension possible if additional time is needed. Fees for copying may be charged but must be reasonable and based on actual costs.

Amendment rights permit patients to request changes to their PHI when they believe information is inaccurate or incomplete. Organizations may deny amendment requests if the information was not created by the organization, is not available for access, or is accurate and complete. Denial letters must explain the basis for denial and inform patients of their right to submit disagreement statements.

Accounting of disclosures requires organizations to track and report PHI disclosures made for certain purposes during the six years prior to the request. This accounting must include disclosure dates, recipients, purposes, and descriptions of information disclosed.

Electronic health record systems should maintain disclosure logs automatically to support these requirements.

Restriction requests allow patients to ask organizations to limit uses or disclosures of their PHI. Organizations are not required to agree to most restriction requests but must honor agreements they accept. However, organizations must honor restriction requests for disclosures to health plans when patients pay out-of-pocket for healthcare services.

Business Associate Requirements

Business associate relationships create extended privacy obligations that require careful management through contracts and ongoing oversight. These relationships exist when external organizations handle PHI on behalf of covered entities, creating shared responsibility for privacy protection.

Business associate agreements (BAAs) establish privacy obligations for external organizations that handle PHI. These agreements must specify permitted uses and disclosures, require appropriate safeguards, prohibit unauthorized uses or disclosures, and establish termination procedures for compliance failures.

Common business associate relationships include billing companies, transcription services, legal counsel, consultants, cloud service providers, and medical equipment vendors. Organizations must identify all business associate relationships and ensure appropriate agreements are in place before PHI sharing begins.

Business associate oversight requires ongoing monitoring of compliance with agreement terms. This includes regular compliance assessments, incident reporting requirements, and audit rights that allow covered entities to verify appropriate privacy protections.

Subcontractor relationships occur when business associates engage other organizations to handle PHI. Business associates must ensure their subcontractors provide appropriate privacy protections through written agreements that mirror BAA requirements.

HIPAA Security Rule Implementation

The Security Rule requires covered entities to implement administrative, physical, and technical safeguards to protect electronic PHI (ePHI). Unlike the Privacy Rule's broad principles, the Security Rule provides specific implementation requirements that organizations must address systematically.

Administrative Safeguards

Administrative safeguards establish the foundation for ePHI protection through policies, procedures, and workforce management. These safeguards address human factors that often represent the weakest links in information security programs.

The Security Officer requirement designates a single individual responsible for developing and implementing security policies and procedures. This person must have sufficient authority and resources to fulfill security responsibilities effectively. Large organizations may have security teams, but accountability must rest with a designated individual.

Workforce training ensures employees understand their security responsibilities and know how to protect ePHI appropriately. Training should cover password management, workstation security, email safety, and incident reporting procedures. Role-specific training addresses unique security challenges for different job functions.

Information system activity review requires regular monitoring of ePHI access and use through audit logs and other monitoring mechanisms. Organizations should establish procedures for log review, anomaly detection, and investigation of suspicious activities. Automated monitoring tools can help identify potential security incidents.

Contingency planning addresses how organizations will protect ePHI during emergencies, system failures, and other disruptions. Contingency plans should include data backup procedures, disaster recovery processes, emergency access procedures, and business continuity planning for critical systems.

Case Example 2: The Premera Blue Cross Cyberattack In 2014, Premera Blue Cross experienced a cyberattack that compromised 11 million patient records. The attack went undetected for several months due to inadequate monitoring systems and insufficient incident response procedures. Premera paid $6.85 million to settle HIPAA violations related to inadequate administrative safeguards, including failure to conduct accurate risk assessments and implement adequate monitoring systems.

Physical Safeguards

Physical safeguards protect computer systems, equipment, and media containing ePHI from unauthorized access, tampering, and theft. These safeguards address traditional security concerns while adapting to modern technology environments.

Facility access controls limit physical access to information systems and equipment. This includes building access controls, server room security, and workstation placement considerations. Organizations should implement appropriate access controls based on facility design and security risks.

Workstation controls govern the functions that can be performed at specific computer workstations and limit access to those workstations to authorized users. This includes physical positioning of workstations, screen placement to prevent unauthorized viewing, and automatic logoff procedures.

Device and media controls address the receipt, removal, and disposal of hardware and electronic media containing ePHI. Organizations must have procedures for sanitizing or destroying media before disposal and maintaining inventory records for equipment containing ePHI.

Technical Safeguards

Technical safeguards use technology to protect ePHI and control access to computer systems containing health information. These safeguards address modern cybersecurity threats while supporting legitimate healthcare operations.

Access control systems ensure that only authorized users can access ePHI and that access is limited to information needed for job functions. This includes user authentication, role-based access controls, and automatic logoff procedures. Multi-factor authentication provides additional security for high-risk access scenarios.

Audit controls monitor and record access to ePHI, creating accountability for information use. Audit logs should capture user identification, access times, actions performed, and information accessed. Regular review of audit logs helps detect unauthorized access and supports incident investigations.

Integrity controls protect ePHI from unauthorized alteration or destruction. This includes version control systems, electronic signatures, and backup procedures that preserve information authenticity. Database integrity controls prevent unauthorized modifications to patient records.

Transmission security protects ePHI during electronic transmission over networks. This includes encryption for data in transit, secure email systems, and virtual private networks for remote access. Network security controls protect against interception and unauthorized access during transmission.

Breach Notification Rule

The Breach Notification Rule requires covered entities to notify patients, the government, and in some cases the media when breaches of unsecured PHI occur. Understanding breach determination processes and notification requirements helps organizations respond appropriately to security incidents.

Breach Determination and Risk Assessment

Breach determination begins with identifying incidents that involve unauthorized acquisition, access, use, or disclosure of PHI. Not every incident constitutes a breach under HIPAA, requiring careful analysis of circumstances and risk factors.

The regulatory definition of breach excludes certain situations including unintentional access by workforce members acting in good faith, inadvertent disclosures between authorized persons at the same organization, and disclosures where information cannot reasonably be retained by unauthorized recipients.

Risk assessment processes evaluate whether incidents create significant risk of financial, reputational, or other harm to individuals. This four-factor analysis considers the nature and extent of PHI involved, the unauthorized person who received PHI, whether PHI was actually acquired or viewed, and the extent to which risk has been mitigated.

Documentation of risk assessments supports breach determination decisions and demonstrates compliance with notification requirements. Risk assessments should be thorough, objective, and completed by individuals with appropriate expertise in privacy and security matters.

Notification Timelines and Requirements

Individual notification must occur within 60 days of breach discovery for incidents affecting 500 or more individuals. Notification may be by first-class mail, email if the individual agreed to electronic communication, or substitute methods if contact information is insufficient or out-of-date.

Media notification is required for breaches affecting 500 or more individuals in a state or jurisdiction. Notification must be provided to prominent media outlets serving the affected area within 60 days of breach discovery. The notification should include the same information provided to individuals.

Government notification to the Office for Civil Rights (OCR) must occur within 60 days for breaches affecting 500 or more individuals. Smaller breaches must be reported annually within 60 days of the end of the calendar year. Electronic reporting through the OCR website is required for all notifications.

Business associate notification obligations require prompt notification to covered entities when breaches occur at business associate organizations. This notification enables covered entities to fulfill their own notification obligations and take appropriate corrective actions.

Case Example 3: The Orlando Health Breach Response In 2016, Orlando Health discovered that an employee had accessed patient records inappropriately over several years, affecting approximately 3,200 patients. The organization conducted a thorough risk assessment, implemented additional security controls, and provided comprehensive notification to affected individuals. Despite the proactive response, Orlando Health paid $85,000 to settle HIPAA violations, demonstrating that even good faith compliance efforts may not prevent penalties when breaches occur.

2025 Updates and Enforcement Focus

HIPAA enforcement has intensified significantly in recent years, with OCR focusing on specific areas that represent the greatest privacy and security risks. Understanding current enforcement priorities helps organizations allocate resources effectively and avoid common compliance pitfalls.

Risk Analysis Enforcement Initiative

OCR has identified risk analysis as a foundational requirement that many organizations fail to implement adequately. Risk analyses must be accurate, thorough, and updated regularly to reflect changing technology and threat environments. Organizations that conduct superficial or outdated risk analyses face significant penalties.

Comprehensive risk analyses evaluate all ePHI within organizational control, including systems, applications, and business processes that create, receive, maintain, or transmit health information. The analysis must consider current and emerging threats, existing safeguards, and residual risks after safeguard implementation.

Risk management processes must address identified vulnerabilities through appropriate corrective actions. Organizations cannot simply identify risks without taking reasonable steps to address them. Risk management should be ongoing rather than a one-time activity, with regular reassessment and safeguard updates.

Documentation requirements for risk analyses include methodology descriptions, findings, and corrective action plans. OCR reviews risk analysis documentation during investigations and audits, using inadequate documentation as evidence of non-compliance.

Cybersecurity Requirements

Cybersecurity threats have evolved dramatically since HIPAA implementation, requiring organizations to adapt their security programs to address sophisticated attack methods. OCR enforcement reflects these changing threats through increased focus on technical safeguards and incident response capabilities.

Ransomware attacks represent a particular concern, as they can simultaneously violate multiple HIPAA requirements including access controls, integrity controls, and contingency planning.

Organizations must implement comprehensive cybersecurity programs that address these evolving threats.

Business associate cybersecurity has become a major enforcement focus, as many breaches originate with third-party vendors. Organizations must evaluate business associate security programs and ensure appropriate contractual protections are in place.

Incident response capabilities affect both breach severity and enforcement outcomes. Organizations with robust incident response programs often experience better enforcement outcomes than those that fail to respond appropriately to security incidents.

Online Tracking Technologies

OCR has increased enforcement attention on online tracking technologies that may impermissibly disclose PHI to technology vendors. Patient portals, appointment scheduling systems, and telehealth platforms often include tracking pixels and analytics tools that share information with third parties.

Cookie management and web analytics require careful evaluation to ensure PHI is not disclosed inappropriately. Organizations must review their websites and online systems to identify tracking technologies and implement appropriate safeguards.

Business associate agreements may be required for technology vendors that receive PHI through tracking pixels or analytics tools. Organizations should evaluate these relationships carefully and implement appropriate contractual protections.

Patient notification about online tracking may be required through privacy notices or other communication mechanisms. Organizations should consider how to inform patients about tracking technologies while maintaining transparency about information sharing practices.

Practical Implementation Strategies

HIPAA compliance requires systematic implementation approaches that address regulatory requirements while supporting healthcare operations. Successful organizations develop comprehensive programs that integrate privacy and security into daily workflows rather than treating them as separate functions.

Risk-based approaches focus resources on areas with greatest privacy and security risks while ensuring adequate coverage of all HIPAA requirements. Organizations should prioritize high-risk activities including research, billing, and marketing while maintaining baseline protections for all PHI handling.

Staff engagement strategies help employees understand why privacy and security matter and how their actions affect patient protection. Training programs should use real-world examples and practical scenarios rather than abstract regulatory concepts.

Technology integration ensures that privacy and security protections are built into information systems rather than added as afterthoughts. Organizations should include HIPAA requirements in system selection, implementation, and maintenance processes.

Continuous improvement approaches treat HIPAA compliance as an ongoing process rather than a one-time achievement. Regular assessments, policy updates, and program enhancements ensure continued effectiveness as threats and technology evolve.

Sustaining Excellence

HIPAA compliance represents more than regulatory obligation—it demonstrates organizational commitment to patient trust and information stewardship. The privacy and security protections required by HIPAA support the therapeutic relationship between patients and healthcare providers while enabling appropriate information sharing for treatment, payment, and healthcare operations.

Effective HIPAA programs balance protection with accessibility, ensuring that privacy and security requirements do not interfere with patient care or legitimate business operations. This balance requires ongoing attention and adjustment as technology evolves and organizational needs change.

The investment in HIPAA compliance pays dividends beyond regulatory compliance through improved information security, enhanced patient trust, and reduced risk of costly breaches. Organizations that view HIPAA as an opportunity to strengthen their information governance rather than just a compliance burden achieve better outcomes for patients, employees, and stakeholders.

Key Takeaways

- HIPAA's three rules create interconnected requirements for privacy, security, and breach notification that must be implemented systematically across healthcare organizations
- Risk analysis serves as the foundation for Security Rule compliance and has become a major enforcement focus requiring regular updates and thorough documentation
- Business associate relationships extend HIPAA obligations beyond covered entities, requiring careful contract management and ongoing oversight
- Breach notification requirements involve complex risk assessments and strict timelines that organizations must prepare to meet before incidents occur
- Current enforcement focuses on cybersecurity, risk analysis adequacy, and online tracking technologies that may inappropriately disclose patient information

Chapter 4: Fraud and Abuse Laws

Healthcare fraud and abuse laws form the backbone of financial integrity in American healthcare, protecting both federal programs and private payers from exploitation. These laws create complex webs of prohibited relationships and activities that can trap unwary healthcare organizations in serious legal violations. Understanding the intricacies of the Stark Law, Anti-Kickback Statute, and False Claims Act is essential for healthcare compliance professionals who must navigate these regulations daily while supporting legitimate business operations.

The Stark Law (Physician Self-Referral)

The Physician Self-Referral Law, commonly known as the Stark Law, prohibits physicians from referring patients for designated health services to entities with which they have financial relationships unless an exception applies. This strict liability statute requires no proof of intent to violate, making technical compliance essential for healthcare organizations.

Designated Health Services and Financial Relationships

Designated Health Services (DHS) encompass eleven categories of services including clinical laboratory services, physical therapy, radiology, radiation therapy, durable medical equipment, parenteral and enteral nutrients, prosthetics, home health services, outpatient prescription drugs, and inpatient and outpatient hospital services. Each category has specific definitions that determine when Stark Law restrictions apply.

The broad definition of hospital services means that most physician referrals to hospitals trigger Stark Law analysis. This includes not only admissions but also outpatient procedures, emergency department visits, and ancillary services provided in hospital settings.

Even referrals within integrated health systems can create Stark Law violations if proper exceptions are not met.

Financial relationships include both ownership interests and compensation arrangements between physicians and entities providing DHS. Ownership interests can be direct or indirect, through equity, debt, or other means. Compensation arrangements include any arrangement involving remuneration between a physician and an entity, regardless of form or purpose.

The "knows or should know" standard applies to both physicians making referrals and entities receiving them. This means that actual knowledge of financial relationships is not required—constructive knowledge based on information reasonably available is sufficient to create liability. Organizations cannot avoid Stark Law compliance by remaining willfully ignorant of physician financial relationships.

Exceptions and Safe Harbors

Stark Law exceptions provide the only means to maintain financial relationships with referring physicians while continuing to receive their referrals. These exceptions have specific requirements that must be met exactly—partial compliance does not provide protection. Common exceptions include employment relationships, personal service arrangements, fair market value transactions, and academic medical center arrangements.

Case Example 1: The Tuomey Healthcare System Case Tuomey Healthcare System in South Carolina entered into employment and service agreements with specialist physicians that included productivity bonuses based on collections from technical component services. The court found these arrangements violated Stark Law because they took into account the volume or value of referrals. Tuomey was required to pay $237 million in penalties, demonstrating how seemingly reasonable physician compensation arrangements can create massive Stark Law liability.

The employment exception protects bona fide employment relationships where physicians are actual employees rather than independent contractors. The compensation must be for identifiable services, consistent with fair market value, not based on referral volume, and commercially reasonable. Part-time employment can qualify, but the relationship must be genuine employment rather than a sham arrangement to avoid Stark Law restrictions.

Personal service arrangements exception covers contracts for physician services including medical director agreements, call coverage arrangements, and consulting relationships. These arrangements must be in writing, specify the services covered, have terms of at least one year, include fair market value compensation not based on referrals, and be commercially reasonable.

Fair market value represents a critical concept throughout Stark Law exceptions. Compensation is considered fair market value if it would be paid in an arm's length transaction for identical services in the same geographic area. Fair market value cannot account for referral volume or value, meaning that compensation above market rates cannot be justified by referral patterns.

Value-Based Arrangement Exceptions

Recent Stark Law changes introduced new exceptions for value-based arrangements that allow financial relationships focused on coordinating and improving patient care while reducing costs. These exceptions recognize the shift toward value-based healthcare while maintaining protections against improper referral relationships.

Value-based enterprises (VBEs) are groups of participants collaborating to achieve value-based purposes including coordinating and managing care, improving quality, or appropriately reducing costs. VBEs can include hospitals, physician practices, and other healthcare providers working together toward common goals.

Value-based arrangements within VBEs can include shared savings distributions, risk-sharing payments, and other financial arrangements that reward improved outcomes or cost reductions. These arrangements must meet specific requirements including written agreements, value-based purposes, and appropriate monitoring and assessment mechanisms.

The exception requirements vary based on financial risk levels within arrangements. Full financial risk arrangements have the most flexibility, while arrangements without downside financial risk have more restrictive requirements including outcome measures and monitoring obligations.

Documentation and Compliance Requirements

Stark Law compliance requires meticulous documentation and ongoing monitoring of physician financial relationships. Organizations must maintain complete records of all arrangements with physicians, including contracts, amendments, and compensation calculations. Documentation must support exception compliance at all times.

Annual reviews of physician arrangements help ensure continued compliance as circumstances change. These reviews should evaluate fair market value, commercial reasonableness, and exception requirements. Changes in physician compensation, services provided, or referral patterns may affect compliance status.

Referral tracking systems help organizations monitor DHS referrals from physicians with financial relationships. These systems should identify prohibited referrals and flag potential compliance issues before they result in claim submissions. Automated tracking reduces compliance burdens while improving accuracy.

Case Example 2: The North Broward Hospital District Settlement
North Broward Hospital District paid $69.5 million to resolve allegations that they paid physicians above fair market value for

services and provided free or below-cost services to physician practices. The government alleged these arrangements were intended to induce referrals in violation of Stark Law. The settlement included a Corporate Integrity Agreement requiring ongoing monitoring of physician arrangements and fair market value assessments.

Anti-Kickback Statute (AKS)

The Anti-Kickback Statute prohibits knowingly and willfully offering, paying, soliciting, or receiving remuneration to induce referrals for services payable by federal healthcare programs. Unlike Stark Law's strict liability standard, AKS requires proof of intent but applies more broadly to all healthcare services and relationships.

Elements of AKS Violations

AKS violations require four elements: remuneration, intent to induce referrals, knowledge that services are payable by federal programs, and willful conduct. Each element must be proven beyond a reasonable doubt in criminal cases, but civil enforcement requires only a preponderance of evidence.

Remuneration includes anything of value transferred between parties, regardless of form or characterization. This encompasses cash payments, reduced rent, free services, entertainment, gifts, and any other benefit provided to influence referral decisions. The breadth of this definition means that many business courtesies can potentially violate AKS.

Intent to induce referrals can be proven through direct evidence such as explicit agreements or communications, but more commonly relies on circumstantial evidence including payment timing, amounts, and relationships to referral patterns. Even if legitimate purposes exist for arrangements, improper intent can still result in violations.

The "one purpose" test means that if one purpose of an arrangement is to induce referrals, AKS may be violated even if other legitimate purposes exist. This creates risk for arrangements with multiple purposes, requiring careful structuring to avoid appearance of improper intent.

Safe Harbors and Protection

AKS safe harbors provide complete protection from prosecution when their requirements are met exactly. Unlike Stark Law exceptions, safe harbor protection is optional—arrangements that do not meet safe harbors can still be legal if they do not violate the underlying statute. However, safe harbor compliance provides certainty and protection from prosecution.

The personal services safe harbor protects legitimate service agreements between healthcare providers and referral sources. Requirements include written agreements, term of at least one year, specific description of services, fair market value compensation set in advance, and compensation not based on referral volume or value.

The employment safe harbor covers bona fide employment relationships where employees provide services for employers. The employment must be genuine rather than a sham arrangement, and compensation must be consistent with fair market value for actual services provided.

Space and equipment rental safe harbors protect legitimate real estate and equipment leasing arrangements. These safe harbors require written agreements, fair market value rent, terms of at least one year, and rent not based on referral volume. The space or equipment must be used exclusively by the lessee during rental periods.

Case Example 3: The Novartis Pharmaceutical Settlement
Novartis paid over $729 million to resolve allegations that they provided kickbacks to healthcare providers through speaker programs, consulting arrangements, and other benefits to induce prescriptions of their medications. The government alleged these programs had little legitimate purpose and were primarily intended to influence

prescribing decisions. The case demonstrates how seemingly legitimate business activities can violate AKS when improper intent is present.

Relationship with False Claims Act

AKS violations create potential False Claims Act liability because claims submitted for services resulting from kickbacks are considered false or fraudulent. This connection means that AKS violations can result in both criminal prosecution and civil monetary penalties under multiple statutes.

The Patient Protection and Affordable Care Act established that claims resulting from AKS violations are automatically false claims for purposes of civil enforcement. This eliminated the need to prove actual knowledge of falsity, making civil enforcement of AKS violations more straightforward for government prosecutors.

Treble damages under the False Claims Act can multiply penalties significantly when AKS violations result in submitted claims. Organizations may face penalties of three times the amount of claims submitted plus civil monetary penalties for each false claim, creating enormous financial exposure.

Recent Enforcement Trends

AKS enforcement has increased significantly in recent years, with government agencies focusing on pharmaceutical companies, device manufacturers, and healthcare providers involved in referral relationships. Enforcement priorities include speaker programs, consulting arrangements, research agreements, and physician ownership interests.

Pharmaceutical speaker programs receive particular scrutiny when they appear designed to influence prescribing rather than provide legitimate education. Enforcement agencies examine speaker selection criteria, program content, venue selection, and relationships between speaker payments and prescribing patterns.

Medical device companies face enforcement for arrangements with physician users including consulting agreements, royalties, and investment opportunities. These arrangements are scrutinized for relationships between financial benefits and device utilization patterns.

False Claims Act (FCA)

The False Claims Act provides the government's primary tool for recovering funds lost to healthcare fraud, allowing both government enforcement and private citizen whistleblower lawsuits. FCA penalties can be devastating for healthcare organizations, making prevention and early detection essential for compliance programs.

Elements of FCA Liability

FCA violations occur when organizations knowingly submit false or fraudulent claims for payment to federal programs. The knowledge standard includes actual knowledge, deliberate ignorance, and reckless disregard for truth or falsity. Organizations cannot avoid liability by remaining willfully ignorant of claim accuracy.

False claims can result from incorrect coding, billing for services not provided, billing for medically unnecessary services, or failing to meet program requirements. The breadth of potential FCA violations means that many compliance failures can result in false claims liability.

Materiality requires that false statements have a natural tendency to influence payment decisions or have the capability of influencing such decisions. This standard focuses on the government's payment practices rather than whether specific claims would have been paid differently.

Scienter requirements mean that organizations must have knowledge of falsity or act with reckless disregard for claim accuracy. Good faith efforts to comply with program requirements can provide protection, but negligent or reckless conduct can still result in liability.

Qui Tam Provisions and Whistleblower Protection

Qui tam provisions allow private citizens with knowledge of false claims to file lawsuits on behalf of the government and share in any recovery. These lawsuits, filed by relators or whistleblowers, comprise the majority of FCA cases and recoveries in healthcare.

Relator requirements include original source knowledge of violations and filing before public disclosure of the alleged violations. Relators must have direct and independent knowledge rather than information derived from public sources or other whistleblower lawsuits.

Government intervention decisions determine whether the Department of Justice will take over qui tam lawsuits or allow relators to proceed independently. Government intervention significantly increases the likelihood of successful resolution and higher recoveries.

Whistleblower protections prohibit retaliation against employees who report suspected FCA violations or participate in investigations. Protected activities include internal reporting, government cooperation, and refusing to participate in unlawful conduct. Retaliation can result in additional penalties and damage awards.

Healthcare-Specific Applications

Healthcare FCA applications encompass virtually every aspect of healthcare operations including billing and coding, medical necessity determinations, quality reporting, and regulatory compliance. The complexity of healthcare regulations creates numerous opportunities for false claims violations.

Upcoding occurs when organizations submit claims for higher-paying services than actually provided or documented. This includes using incorrect procedure codes, diagnosis codes that do not reflect patient conditions, or modifier usage that inappropriately increases reimbursement.

Medical necessity violations involve billing for services that are not reasonable and necessary for patient diagnosis or treatment. Organizations must maintain documentation supporting medical necessity determinations and ensure that services meet program coverage requirements.

Quality reporting false claims can result from inaccurate data submission for pay-for-performance programs, meaningful use attestations, or other quality initiatives. Organizations must ensure accuracy of quality data and maintain supporting documentation for reported metrics.

Damages and Penalties

FCA penalties include civil monetary penalties of $13,508 to $27,018 per false claim (adjusted annually for inflation) plus treble damages equal to three times the government's actual damages. These penalties can accumulate rapidly when violations involve multiple claims over extended periods.

Calculation of damages includes both direct payments received and consequential damages such as interest and administrative costs. In healthcare cases, damages often equal the full amount of payments received for false claims rather than just the overpayment amount.

Settlement negotiations often result in reduced penalties in exchange for cooperation, corrective actions, and admission of liability. However, settlement amounts can still be substantial, particularly when violations involve systematic problems or extended time periods.

Practical Application and Compliance Strategies

Healthcare organizations need systematic approaches to fraud and abuse compliance that address the complexity and overlap of these laws while supporting legitimate business operations. Effective compliance programs integrate legal requirements into operational processes rather than treating them as separate functions.

Financial Arrangement Reviews

Systematic review processes help organizations identify and evaluate financial relationships with physicians and other referral sources. These reviews should occur before arrangements begin, during annual assessments, and when circumstances change significantly.

Review criteria should address both legal requirements and business purposes for arrangements. Legal analysis must consider applicable exceptions and safe harbors while business analysis should evaluate commercial reasonableness and strategic alignment with organizational goals.

Documentation requirements include written contracts, fair market value assessments, legal analyses, and approval records. This documentation supports compliance decisions and provides evidence of good faith efforts to comply with legal requirements.

Fair Market Value Determinations

Fair market value assessments provide critical protection for physician financial arrangements by ensuring compensation reflects legitimate business purposes rather than referral relationships. These assessments should be objective, well-documented, and based on reliable market data.

Valuation methodologies should consider comparable transactions, professional surveys, and economic analysis of services provided.

Multiple data sources strengthen valuation conclusions and provide more defensible fair market value determinations.

Independent assessments by qualified valuation professionals provide additional credibility and objectivity for significant arrangements. These assessments should be updated periodically to ensure continued fair market value compliance as market conditions change.

Self-Disclosure Protocols

Self-disclosure programs allow organizations to report potential violations voluntarily in exchange for reduced penalties and negotiated resolution agreements. These programs can provide significant benefits when violations are discovered internally.

Evaluation criteria for self-disclosure include violation severity, cooperation level, corrective actions implemented, and organizational compliance history. Organizations should consider legal counsel guidance when evaluating disclosure decisions.

Disclosure documentation should be thorough, accurate, and demonstrate good faith efforts to identify and correct violations. Incomplete or inaccurate disclosures can result in worse outcomes than voluntary disclosure might otherwise achieve.

Building Sustainable Compliance

Fraud and abuse compliance requires ongoing vigilance and systematic processes that evolve with changing regulations and business practices. Organizations that view these laws as constraints rather than protections miss opportunities to strengthen their operations while avoiding legal pitfalls.

The complexity of these laws should not paralyze healthcare organizations or prevent legitimate business relationships. Instead, understanding fraud and abuse requirements enables organizations to structure arrangements properly, document decisions thoroughly, and monitor compliance effectively.

Successful fraud and abuse compliance programs balance legal protection with operational efficiency, ensuring that compliance requirements support rather than hinder healthcare delivery. This balance requires ongoing education, regular assessment, and commitment to ethical business practices throughout the organization.

Key Takeaways

- Stark Law creates strict liability for physician self-referral violations, requiring exact compliance with exception requirements rather than good faith efforts
- Anti-Kickback Statute requires proof of intent but applies broadly to all federal program services, with safe harbors providing optional but complete protection
- False Claims Act creates potential treble damages and massive penalties for claims resulting from fraud and abuse violations, with whistleblower lawsuits driving most enforcement
- Fair market value determinations and proper documentation provide essential protection for physician financial arrangements across all three statutes
- Self-disclosure programs offer opportunities for reduced penalties when violations are discovered, but require careful evaluation and thorough documentation

Chapter 5: CMS Requirements and Value-Based Care

The Centers for Medicare & Medicaid Services represents the largest single payer in American healthcare, making its requirements and programs central to healthcare compliance. CMS regulations extend far beyond simple billing rules to encompass quality reporting, patient safety, and emerging value-based care models that reward outcomes over volume. Understanding these evolving requirements is essential for healthcare organizations seeking to maintain compliance while thriving in an increasingly complex regulatory environment.

Conditions of Participation

The Conditions of Participation (CoPs) establish the foundational requirements that healthcare organizations must meet to participate in Medicare and Medicaid programs. These standards address organizational structure, patient care processes, and quality assurance mechanisms that ensure safe, effective healthcare delivery.

Hospital Requirements and Governing Structure

Hospital CoPs require formal governing structures with clear accountability for organizational compliance and quality of care. The governing body must ensure appropriate medical staff oversight, maintain effective quality assurance programs, and provide adequate resources for safe patient care. These requirements create legal obligations for board members and senior executives who may face personal liability for compliance failures.

Medical staff requirements establish credentialing standards, peer review processes, and clinical oversight mechanisms. Hospitals must verify physician qualifications, monitor clinical performance, and take corrective action when quality problems are identified. The medical staff must have organized leadership, established bylaws, and regular meetings to address clinical issues.

Nursing services requirements mandate adequate staffing levels, appropriate supervision, and competency-based care delivery. The director of nursing must be a registered nurse with appropriate qualifications and authority to manage nursing operations. Staffing levels must be adequate to meet patient care needs at all times, with appropriate skill mix and supervision.

Patient rights requirements ensure that hospitals respect patient autonomy, provide appropriate information, and protect patient dignity throughout the care process. Patients must receive information about their condition, treatment options, and rights to participate in care decisions. Hospitals must have processes for addressing patient complaints and grievances.

Case Example 1: The West Boca Medical Center Investigation In 2019, CMS terminated West Boca Medical Center from Medicare participation due to immediate jeopardy violations including inadequate infection control, medication management failures, and patient safety problems. The hospital was required to develop comprehensive corrective action plans and demonstrate sustained compliance before Medicare participation could be restored. This case demonstrates how CoP violations can threaten organizational survival through loss of Medicare participation.

Quality Reporting Obligations

Quality reporting requirements mandate that hospitals collect, analyze, and report performance data for various quality measures. These reports inform public reporting systems, payment adjustments, and regulatory oversight activities. Failure to meet reporting requirements can result in payment reductions and increased regulatory scrutiny.

The Hospital Inpatient Quality Reporting (IQR) program requires reporting on clinical process measures, outcome measures, and patient experience scores. Hospitals must submit data through established reporting systems, meet data accuracy standards, and comply with public reporting requirements. Payment reductions apply to hospitals that fail to meet reporting requirements.

Hospital Outpatient Quality Reporting (OQR) program requirements address ambulatory surgery centers and hospital outpatient departments. These requirements include clinical quality measures, patient safety indicators, and efficiency measures. Reporting requirements vary based on facility type and services provided.

Electronic Clinical Quality Measures (eCQMs) represent the future of quality reporting, using electronic health record data to calculate performance automatically. Hospitals must ensure their EHR systems can capture required data elements and generate accurate quality measure reports. The transition to eCQMs requires significant information technology investments and workflow modifications.

Patient Safety Measures

Patient safety requirements address infection control, medication management, and other clinical processes that affect patient outcomes. These requirements are based on evidence-based practices and are regularly updated to reflect new knowledge about patient safety risks.

Infection prevention and control requirements mandate comprehensive programs to prevent healthcare-associated infections. Hospitals must have infection preventionists, surveillance systems, and evidence-based protocols for preventing common infections. Recent updates address emerging pathogens and antibiotic-resistant organisms.

Medication management requirements ensure safe prescribing, dispensing, and administration of medications throughout the hospital. Pharmacist oversight is required for medication systems, and hospitals must have processes to prevent medication errors and adverse drug events.

Emergency preparedness requirements mandate comprehensive planning for natural disasters, disease outbreaks, and other emergencies that could affect hospital operations. Hospitals must have written emergency plans, conduct regular drills, and coordinate with community emergency response systems.

Medicare and Medicaid Compliance

Medicare and Medicaid compliance extends beyond the Conditions of Participation to include enrollment requirements, billing accuracy, and program integrity measures. These requirements affect all healthcare providers participating in federal programs, from individual physicians to large health systems.

Enrollment and Revalidation

Provider enrollment establishes the legal relationship between healthcare organizations and CMS, creating the foundation for payment and oversight. Enrollment requirements include background checks, site visits, and ongoing monitoring to ensure continued eligibility for program participation.

Revalidation processes require periodic review of provider eligibility and compliance with program requirements. Providers must submit updated information, undergo background checks, and demonstrate continued compliance with enrollment standards. Failure to complete revalidation can result in deactivation from Medicare programs.

Change of ownership transactions require advance notification to CMS and may trigger new enrollment requirements. These transactions can affect existing Medicare agreements, payment arrangements, and compliance obligations. Organizations must carefully manage change of ownership processes to avoid disruption of Medicare participation.

Provider agreement requirements establish the contractual relationship between providers and CMS, creating legally binding obligations for program compliance. These agreements incorporate all applicable regulations, conditions of participation, and payment rules. Violation of provider agreement terms can result in termination from Medicare programs.

Billing and Coding Compliance

Accurate billing and coding form the foundation of Medicare and Medicaid compliance, requiring systematic processes to ensure claims reflect services actually provided to program beneficiaries. Billing errors can result in overpayments that must be returned to the government, even when errors are unintentional.

Documentation requirements support billing accuracy by providing evidence of services rendered and medical necessity determinations. Documentation must be complete, accurate, and available for audit review. Physicians and other healthcare providers must personally document their services and cannot rely solely on templates or copied information.

Medical necessity standards require that services be reasonable and necessary for diagnosis or treatment of patient conditions. Providers must maintain documentation supporting medical necessity determinations and ensure that services meet Medicare coverage criteria. Local coverage determinations and national coverage decisions provide guidance on specific coverage requirements.

Case Example 2: The Halifax Health System Settlement Halifax Health System paid $85 million to resolve allegations of Medicare billing fraud including unnecessary cardiac procedures, emergency department services, and observation stays. The government alleged that the organization systematically billed for services that were not medically necessary or were not provided as billed. The settlement included extensive monitoring requirements and corrective action obligations, demonstrating the serious consequences of billing compliance failures.

Overpayment Obligations

The 60-day overpayment rule requires healthcare organizations to identify, investigate, and return overpayments within 60 days of identification. This obligation applies to all Medicare and Medicaid overpayments, regardless of cause or amount. Failure to return overpayments can result in False Claims Act liability and other penalties.

Identification of overpayments can occur through internal auditing, government audits, whistleblower reports, or other sources. Organizations must have processes to evaluate potential overpayments and determine appropriate responses. Good faith investigations and prompt repayment provide some protection from penalties.

Repayment processes must comply with specific CMS requirements including proper documentation, interest calculations, and reporting mechanisms. Organizations should maintain records of overpayment identification, investigation activities, and repayment actions to demonstrate compliance with 60-day rule requirements.

Value-Based Care Programs

Value-based care represents a fundamental shift from fee-for-service payment toward arrangements that reward quality outcomes and cost efficiency. These programs create new compliance challenges while offering opportunities for increased revenue and improved patient care.

Medicare Shared Savings Program (MSSP)

The Medicare Shared Savings Program enables healthcare organizations to form Accountable Care Organizations (ACOs) that accept responsibility for quality and cost outcomes for defined patient populations. ACOs that achieve quality benchmarks and cost savings share in the savings generated.

ACO formation requires legal entity creation, participant agreements, and governance structures that ensure accountability for patient outcomes. ACOs must have adequate financial resources, management capabilities, and clinical integration to manage population health effectively.

Quality reporting requirements for ACOs include 33 quality measures addressing patient safety, clinical care, patient experience, and population health. ACOs must achieve minimum performance standards on quality measures to be eligible for shared savings payments. Quality reporting uses both claims-based measures and consumer assessment surveys.

Financial reconciliation processes determine shared savings payments based on actual versus expected costs for ACO beneficiaries. These calculations consider quality performance, beneficiary attribution, and risk adjustment factors. ACOs with downside financial risk can also be liable for shared losses when costs exceed benchmarks.

Case Example 3: The Geisinger Health System ACO Success
Geisinger Health System became one of the most successful ACOs in the Medicare Shared Savings Program, achieving both quality goals and cost savings consistently since program inception. Their integrated delivery model, robust data analytics, and focus on care coordination enabled them to reduce hospital readmissions by 18% while improving patient satisfaction scores. Geisinger earned over $100 million in shared savings payments during their first five years of participation, demonstrating how effective population health management can create financial rewards under value-based payment models.

MIPS and Advanced Payment Models

The Merit-based Incentive Payment System (MIPS) applies to most physicians participating in Medicare, adjusting payments based on performance in four categories: quality, cost, improvement activities, and promoting interoperability. MIPS creates both opportunities for bonus payments and risks of payment reductions based on comparative performance.

Quality performance category accounts for 30% of the MIPS composite score and includes reporting on six quality measures relevant to clinical specialty and practice setting. Physicians can report through multiple mechanisms including claims, qualified clinical data registries, electronic health records, and web interfaces. Benchmark comparisons determine performance scores relative to other providers.

Cost performance category evaluates spending per beneficiary and episode-based cost measures where applicable. CMS calculates cost scores using claims data, comparing actual spending to expected costs based on beneficiary characteristics and geographic factors. This category accounts for 30% of the composite score and increases physician accountability for healthcare spending.

Improvement activities category rewards participation in activities that improve clinical practice, care coordination, beneficiary engagement, or patient safety. Examples include participation in specialty society quality improvement programs, use of decision support tools, and implementation of fall prevention programs. This category provides 15% of the composite score.

Promoting interoperability category replaces the meaningful use program for MIPS participants, focusing on electronic health record functionality that supports care coordination and patient engagement. Requirements include electronic prescribing, health information exchange, and patient electronic access to health information. This category contributes 25% of the composite score.

Advanced Payment Models (APMs) provide alternative pathways for physicians who participate in qualifying value-based arrangements. APM participants can earn 5% bonus payments and avoid MIPS reporting requirements when they meet minimum participation thresholds. Qualifying APMs include comprehensive primary care payment models, bundled payment arrangements, and shared savings programs with downside financial risk.

Data Completeness and Accuracy Requirements

Value-based care programs depend on accurate, complete data to measure performance and calculate payments appropriately. Healthcare organizations must ensure their data collection, validation, and reporting processes meet program requirements while supporting clinical decision-making.

Electronic health record integration becomes essential for efficient data collection and reporting in value-based programs. Organizations must configure their EHR systems to capture required data elements, generate performance reports, and support quality improvement activities. EHR vendors play critical roles in enabling value-based care participation through system functionality and reporting capabilities.

Data validation processes ensure accuracy of performance measures and payment calculations. Organizations should implement regular data audits, reconciliation procedures, and quality checks to identify and correct errors before submitting reports to CMS. Inaccurate data can result in payment adjustments, penalties, and compliance issues.

2025 Regulatory Updates

CMS continues to evolve its payment systems and regulatory requirements, with 2025 bringing significant changes that affect healthcare organizations across the country. These updates reflect lessons learned from existing programs while addressing emerging healthcare challenges and technological capabilities.

TEAM Model Implementation

The Transforming Episode Accountability Model (TEAM) represents CMS's latest approach to bundled payments, building on lessons learned from previous episode payment initiatives. TEAM focuses on high-volume, high-cost procedures including joint replacements, cardiac procedures, and other surgical interventions.

Episode definition improvements in TEAM address previous program limitations by including post-acute care services, readmissions, and complications in episode calculations. This comprehensive approach encourages providers to manage entire episodes of care rather than just the initial procedure.

Quality requirements in TEAM include both outcome measures and process measures relevant to episode types. Providers must achieve minimum quality thresholds to earn reconciliation payments and avoid penalties. Quality measures address complications, readmissions, patient experience, and functional outcomes.

Participation requirements allow both mandatory and voluntary participation depending on geographic location and provider characteristics. Mandatory participants include hospitals in selected metropolitan areas for covered episode types. Voluntary participants can join TEAM in other areas or for additional episode types.

eCQM Transition Requirements

The transition from chart-abstracted quality measures to electronic Clinical Quality Measures (eCQMs) represents a fundamental shift in quality reporting methodology. This transition requires significant investments in health information technology while promising more efficient and accurate quality measurement.

Implementation timelines vary by provider type and reporting program, with hospitals leading the transition followed by physician practices and other provider types. CMS provides implementation support including technical assistance, testing opportunities, and feedback reports to help providers succeed with eCQM reporting.

Technical requirements for eCQM reporting include certified EHR technology, data extraction capabilities, and quality reporting modules. Providers must ensure their systems can calculate quality measures accurately and generate required reports in specified formats. Vendor partnerships become critical for successful eCQM implementation.

Data accuracy standards for eCQMs require validation procedures, audit trails, and documentation of measure calculations. Providers must demonstrate that their eCQM data accurately reflects patient care and clinical outcomes. CMS may audit eCQM submissions and require corrective actions for accuracy problems.

Interoperability Rules and Information Blocking

Information blocking regulations prohibit practices that interfere with the access, exchange, or use of electronic health information. These rules apply to healthcare providers using certified health IT and create compliance obligations for organizations and individual practitioners.

Exception criteria allow certain practices that might otherwise constitute information blocking when they serve legitimate purposes such as protecting patient safety, privacy, or security. Providers must understand these exceptions and document their application when restricting information access or exchange.

Patient access requirements mandate that healthcare providers give patients electronic access to their health information without special effort. This includes structured data from EHRs that must be available through patient portals or mobile applications. Providers cannot charge fees for electronic access to patient information.

Third-party application requirements enable patients to direct their health information to applications of their choice. Providers must support standardized APIs that allow approved applications to access patient data securely. This requirement supports patient control over health information while enabling innovation in health technology.

Practical Applications and Implementation Strategies

Successfully navigating CMS requirements and value-based care programs requires systematic approaches that integrate regulatory compliance with clinical and operational excellence. Organizations must balance multiple, sometimes competing requirements while maintaining focus on patient care quality and financial sustainability.

Quality Measure Selection and Performance

Strategic selection of quality measures allows organizations to focus improvement efforts on areas that align with clinical priorities and organizational strengths. While some measures are mandatory, voluntary reporting programs often provide choice among multiple measures within categories.

Performance improvement initiatives should address root causes of quality gaps rather than just improving reported scores. Effective initiatives engage clinical staff, use data to drive decision-making, and implement evidence-based practices that improve patient outcomes.

Benchmark analysis helps organizations understand their performance relative to peers and identify opportunities for improvement. CMS provides benchmark data for most quality measures, enabling organizations to assess their competitive position and set realistic improvement targets.

Care Coordination and Population Health

Value-based care success requires systematic approaches to care coordination that ensure patients receive appropriate services across multiple providers and settings. This includes primary care management, specialist coordination, and transitions of care between settings.

Population health management involves identifying high-risk patients, managing chronic conditions, and preventing avoidable complications and hospitalizations. Successful organizations use predictive analytics, care management protocols, and patient engagement strategies to improve population outcomes.

Regulatory Change Management

CMS regulations change frequently, requiring systematic processes to identify, evaluate, and implement regulatory updates. Organizations should designate responsible individuals, establish review procedures, and maintain implementation timelines for regulatory changes.

Communication strategies ensure that regulatory changes reach appropriate staff members and are implemented consistently across the organization. This includes policy updates, training programs, and performance monitoring to ensure sustained compliance with new requirements.

Advancing Healthcare Excellence

CMS requirements and value-based care programs represent more than regulatory compliance—they embody the healthcare system's evolution toward better outcomes, improved efficiency, and enhanced patient experience. Organizations that embrace these changes as opportunities rather than burdens position themselves for long-term success in an evolving healthcare environment.

The complexity of these programs requires sustained commitment to learning, adaptation, and improvement. Healthcare organizations must invest in technology, training, and process improvement while maintaining focus on their fundamental mission of providing excellent patient care.

Success in value-based care ultimately depends on the same factors that drive clinical excellence: systematic processes, engaged staff, continuous improvement, and unwavering commitment to patient welfare. These programs provide frameworks and incentives for achieving these goals while creating financial sustainability for healthcare organizations.

Key Takeaways

- Conditions of Participation establish foundational requirements for Medicare participation, with violations potentially threatening organizational survival through program termination
- Quality reporting requirements continue expanding toward electronic measures that promise greater efficiency but require significant health information technology investments
- Value-based care programs reward quality outcomes and cost efficiency, requiring new capabilities in population health management and care coordination
- The 60-day overpayment rule creates strict obligations for identifying and returning Medicare overpayments, with False Claims Act liability for non-compliance
- 2025 regulatory updates focus on bundled payments, electronic quality measures, and information sharing that support coordinated, patient-centered care

Chapter 6: Additional Regulatory Requirements

Healthcare organizations operate within a complex web of federal, state, and international regulations that extend far beyond the primary compliance frameworks of HIPAA and fraud prevention. These additional requirements address medical devices, emergency care obligations, government oversight agreements, and emerging global privacy standards that increasingly affect healthcare operations. Understanding these diverse regulatory demands helps compliance professionals build comprehensive programs that address all applicable requirements while supporting organizational mission and operations.

FDA Regulations for Healthcare

The Food and Drug Administration plays a critical role in healthcare regulation through oversight of medical devices, pharmaceuticals, and clinical research. Healthcare organizations must understand FDA requirements that affect device usage, drug administration, and research activities conducted within their facilities.

Medical Device Regulations and Quality Systems

Medical device regulations establish safety and effectiveness standards for devices used in healthcare settings, from simple tongue depressors to complex robotic surgical systems. Healthcare organizations must ensure they use only FDA-approved or cleared devices while maintaining appropriate records and reporting adverse events.

Device classification systems categorize medical devices into Class I, II, or III based on risk levels and regulatory requirements. Class I devices have the lowest risk and minimal regulatory requirements, while Class III devices require premarket approval and extensive clinical data. Healthcare organizations must understand device classifications to ensure appropriate procurement and use.

Quality System Regulation (QSR) requirements apply to device manufacturers but affect healthcare organizations that modify devices or create custom solutions. Organizations that make significant modifications to devices may trigger manufacturing regulations and associated quality system requirements.

Medical Device Reporting (MDR) obligations require healthcare organizations to report device-related deaths, serious injuries, and malfunctions to both FDA and device manufacturers. Reports must be submitted within specific timeframes using standardized forms. Failure to report can result in FDA enforcement actions and patient safety risks.

Case Example 1: The Medtronic Infusion Pump Recall In 2019, Medtronic recalled certain MiniMed insulin pumps due to cybersecurity vulnerabilities that could allow unauthorized access to pump controls. Healthcare organizations using these devices had to implement risk mitigation strategies, notify patients, and coordinate device replacements. The recall highlighted how healthcare organizations must monitor device safety communications and respond appropriately to protect patient safety while maintaining continuity of care.

Drug Approval and Marketing Oversight

FDA drug regulations affect healthcare organizations through formulary decisions, off-label use policies, and adverse event reporting requirements. Organizations must ensure appropriate drug procurement, administration, and monitoring while complying with FDA requirements.

Formulary management involves selecting appropriate medications for organizational use based on safety, efficacy, and cost considerations. Pharmacy and therapeutics committees typically oversee formulary decisions and must consider FDA-approved indications, contraindications, and safety communications.

Off-label prescribing allows physicians to use FDA-approved drugs for indications other than those specifically approved by the agency. While legal, off-label use requires careful consideration of available evidence, patient-specific factors, and potential liability issues. Organizations should have policies addressing off-label use decisions and documentation requirements.

Adverse event reporting requirements mandate that healthcare organizations report serious adverse drug events to FDA through the MedWatch system. Healthcare professionals must report deaths, life-threatening events, hospitalizations, disabilities, congenital anomalies, and events requiring intervention to prevent permanent impairment.

Clinical Trial Compliance

Healthcare organizations conducting clinical research must comply with FDA regulations governing investigational drug and device studies. These requirements address study design, participant protection, data integrity, and regulatory reporting.

Good Clinical Practice (GCP) standards establish requirements for clinical trial conduct, including investigator qualifications, protocol adherence, informed consent procedures, and data management. Healthcare organizations must ensure their research programs meet GCP standards to protect participants and maintain scientific integrity.

Institutional Review Board (IRB) oversight ensures that research studies meet ethical standards and protect participant welfare. Healthcare organizations must have appropriate IRB processes, whether through internal boards or external arrangements. IRB approval is required before beginning research involving human subjects.

Investigational Device Exemption (IDE) and Investigational New Drug (IND) applications enable research use of unapproved devices and drugs. Healthcare organizations participating in such studies must comply with specific requirements including participant monitoring, adverse event reporting, and regulatory communications.

EMTALA Compliance

The Emergency Medical Treatment and Labor Act creates specific obligations for hospitals with emergency departments, requiring medical screening examinations and appropriate treatment or transfer for patients with emergency medical conditions. EMTALA violations can result in significant penalties and potential exclusion from Medicare programs.

Medical Screening Examination Requirements

Medical screening examinations must be provided to all individuals who come to hospital emergency departments seeking treatment, regardless of ability to pay or insurance status. These examinations must be conducted by qualified medical personnel using appropriate resources to determine whether emergency medical conditions exist.

Qualified medical personnel typically include physicians, physician assistants, or nurse practitioners with appropriate training and hospital privileges. The hospital medical staff must designate which personnel are qualified to conduct screening examinations and ensure appropriate supervision and oversight.

Screening examination standards must be uniform for all patients presenting with similar symptoms, regardless of insurance status, ability to pay, or other non-medical factors. Hospitals cannot have different screening protocols based on patient characteristics that might suggest inability to pay for services.

Documentation requirements for screening examinations must demonstrate that appropriate evaluation occurred and support any determination regarding the presence or absence of emergency medical conditions. Documentation should include patient presentation, examination findings, diagnostic studies performed, and clinical decision-making.

Stabilization and Transfer Obligations

Stabilization requirements mandate that hospitals provide appropriate treatment to eliminate emergency medical conditions or ensure that conditions will not deteriorate during transfer to another facility. Patients cannot be discharged or transferred until emergency medical conditions are stabilized.

Emergency medical conditions include situations requiring immediate medical attention to prevent serious jeopardy to patient health, serious impairment to bodily functions, or serious dysfunction of body organs. The definition is broad and includes conditions that might not be immediately life-threatening but require prompt attention.

Transfer requirements apply when hospitals cannot provide appropriate treatment for emergency medical conditions or when patients request transfers to other facilities. Transfers must meet specific requirements including physician certification, receiving hospital acceptance, and appropriate medical personnel and equipment during transport.

Case Example 2: The Riverside Regional Medical Center Settlement Riverside Regional Medical Center paid $100,000 to settle EMTALA allegations involving failure to provide appropriate screening examinations and stabilizing treatment to emergency department patients. The investigation found that the hospital discharged patients with emergency medical conditions before providing adequate treatment and failed to document appropriate screening examinations. The settlement required implementation of corrective actions including staff training and policy updates.

On-Call Physician Requirements

On-call physician obligations ensure that hospitals have appropriate specialist coverage available to treat emergency medical conditions that require specialized care. Hospitals must maintain on-call lists and ensure physicians respond appropriately when called.

On-call list requirements mandate that hospitals maintain lists of physicians available to provide specialized treatment for emergency medical conditions. These lists must cover services that the hospital provides and ensure adequate coverage for anticipated patient needs.

Response obligations require on-call physicians to respond promptly when contacted about patients with emergency medical conditions requiring their expertise. Physicians cannot refuse to see patients based on insurance status, ability to pay, or other non-medical factors.

Documentation of on-call arrangements and physician responses helps demonstrate EMTALA compliance and provides evidence of appropriate care coordination. Hospitals should maintain records of on-call schedules, physician contacts, and response times.

Corporate Integrity Agreements

Corporate Integrity Agreements represent negotiated settlements between healthcare organizations and the Office of Inspector General that establish ongoing compliance obligations following fraud and abuse violations. These agreements create extensive monitoring and reporting requirements designed to prevent future violations.

CIA Structure and Requirements

Corporate Integrity Agreements typically include multiple components addressing the specific compliance failures that led to the underlying violations. Common elements include compliance program requirements, training obligations, monitoring activities, and reporting requirements.

Compliance program elements often require organizations to establish or enhance existing compliance programs with specific structural requirements including dedicated compliance officers, compliance committees, and board oversight. Organizations may need to modify reporting relationships or resource allocations to meet CIA requirements.

Training requirements mandate comprehensive education programs for employees, management, and board members addressing specific compliance risks identified during the investigation. Training must be tailored to job functions and updated regularly to address emerging risks.

Monitoring obligations require organizations to conduct regular assessments of compliance with specific legal requirements, often including statistical sampling of claims or transactions. Independent review organizations may be required to conduct some monitoring activities.

Reporting requirements mandate regular submissions to OIG describing compliance activities, monitoring results, and corrective actions implemented. These reports must be signed by senior executives who attest to their accuracy and completeness.

Case Example 3: The Tufts Medical Center CIA Implementation
Tufts Medical Center entered into a comprehensive CIA following allegations of inappropriate emergency department billing and observation services. The five-year agreement required implementation of enhanced compliance monitoring, claims review procedures, and staff training programs. The organization used the CIA requirements as an opportunity to strengthen their overall compliance program, ultimately achieving better outcomes than required minimums and building sustainable compliance processes.

Implementation Strategies and Best Practices

Successful CIA implementation requires systematic project management approaches that address multiple requirements simultaneously while maintaining focus on patient care and organizational operations. Organizations should establish dedicated implementation teams with clear accountability and adequate resources.

Project management methodologies help organizations track CIA requirements, implementation timelines, and deliverable deadlines. Regular progress reporting to senior leadership ensures adequate attention and resource allocation for CIA compliance.

Integration with existing compliance programs prevents duplication of effort while ensuring CIA requirements enhance rather than replace effective existing processes. Organizations should evaluate current programs against CIA requirements and identify opportunities for synergy.

Monitoring and Reporting Excellence

Effective monitoring programs under CIAs focus on continuous improvement rather than just meeting minimum requirements. Organizations that exceed CIA minimums often achieve better long-term compliance outcomes and demonstrate genuine commitment to ethical practices.

Data collection and analysis capabilities must support CIA monitoring requirements while providing meaningful information for organizational decision-making. Investment in data analytics platforms often pays dividends beyond CIA compliance through improved operational insights.

Communication strategies ensure that CIA requirements are understood throughout the organization and integrated into daily operations rather than treated as separate compliance functions. Regular communication reinforces the importance of CIA compliance and recognizes achievements.

State and International Requirements

Healthcare organizations increasingly operate across state boundaries and face international regulatory requirements that add complexity to compliance programs. Understanding these diverse requirements helps organizations avoid violations while expanding their reach and capabilities.

Multi-State Compliance Challenges

Multi-state healthcare operations face varying regulatory requirements that can create conflicting obligations and increase compliance complexity. Organizations must understand state-specific requirements while maintaining consistent quality and ethical standards across all locations.

Professional licensure requirements vary significantly among states, affecting physician practice privileges, nursing scope of practice, and other healthcare professions. Organizations must ensure appropriate licensure for all practitioners and understand practice limitations in different jurisdictions.

Telehealth regulations differ among states regarding licensure requirements, practice standards, and reimbursement policies. Organizations providing telehealth services across state lines must comply with requirements in both originating and destination states.

Privacy and security laws at the state level may impose requirements beyond HIPAA, including breach notification timelines, consent requirements, and patient access rights. Organizations must understand state-specific requirements and implement policies that meet the most stringent applicable standards.

GDPR and International Privacy Requirements

The General Data Protection Regulation affects healthcare organizations that process personal data of European Union residents, creating compliance obligations that extend far beyond traditional healthcare privacy requirements. GDPR's extraterritorial reach means that many US healthcare organizations must comply with European privacy standards.

Personal data under GDPR includes any information relating to identified or identifiable individuals, encompassing health information, contact details, and other identifiers. Health data receives special protection as sensitive personal data with enhanced requirements for processing and disclosure.

Lawful basis requirements mandate that organizations have appropriate legal justification for processing personal data, including health information. Healthcare organizations typically rely on consent, legitimate interests, or public interest grounds, each with specific requirements and limitations.

Individual rights under GDPR include access, rectification, erasure, portability, and objection rights that go beyond HIPAA requirements. Healthcare organizations must have processes to honor these rights while balancing healthcare-specific considerations.

Case Example 4: The International Hospital Corporation GDPR Compliance International Hospital Corporation operates facilities in both the United States and Europe, requiring compliance with both HIPAA and GDPR standards. The organization implemented unified privacy policies that meet both frameworks' requirements, established dual consent processes for research activities, and created data processing agreements that satisfy both regulatory schemes. This approach enabled consistent operations while meeting diverse regulatory requirements.

Data Transfer and Processing Agreements

International data transfers require appropriate safeguards under GDPR, including adequacy decisions, standard contractual clauses, or other approved mechanisms. Healthcare organizations must evaluate their data flows and implement appropriate protections for cross-border transfers.

Vendor management becomes more complex under international privacy requirements, requiring evaluation of data processing activities, contractual protections, and ongoing oversight. Organizations must ensure their vendors meet applicable privacy standards regardless of location.

Cross-Border Research and Clinical Trials

International research collaborations must comply with multiple regulatory frameworks simultaneously, creating complex compliance challenges. Organizations must understand requirements in all participating jurisdictions while maintaining scientific integrity and participant protection.

Good Clinical Practice standards vary among countries, requiring organizations to understand specific requirements for documentation, monitoring, and reporting in each jurisdiction. Harmonized standards help reduce complexity but do not eliminate jurisdictional differences.

Building Comprehensive Compliance Programs

Managing diverse regulatory requirements requires systematic approaches that identify applicable regulations, assess compliance status, and implement effective controls. Successful organizations develop integrated compliance programs that address multiple requirements efficiently while maintaining focus on patient care and safety.

Regulatory Inventory and Assessment

Comprehensive regulatory inventories help organizations identify all applicable requirements and assess current compliance status. These inventories should address federal, state, and international regulations affecting organizational operations.

Risk assessment processes should evaluate the likelihood and impact of violations across different regulatory domains, enabling organizations to prioritize compliance efforts and allocate resources effectively. Regular updates ensure assessments remain current with changing operations and regulations.

Integrated Monitoring and Reporting

Monitoring programs that address multiple regulatory requirements simultaneously reduce administrative burden while improving compliance effectiveness. Organizations should look for opportunities to integrate monitoring activities across different regulatory domains.

Reporting systems should consolidate information from multiple compliance areas while meeting specific requirements for different regulatory authorities. Dashboard approaches can provide senior leadership with comprehensive compliance overviews.

Embracing Regulatory Diversity

The diversity of healthcare regulations reflects the complexity of modern healthcare delivery and the multiple stakeholders with legitimate interests in healthcare operations. Rather than viewing this diversity as a burden, successful organizations recognize that comprehensive compliance programs create competitive advantages through improved quality, reduced risk, and enhanced reputation.

Effective management of diverse regulatory requirements requires commitment to continuous learning, systematic processes, and collaborative approaches that engage subject matter experts across the organization. This investment in compliance infrastructure pays dividends through reduced violations, improved operations, and enhanced patient care.

The regulatory environment will continue evolving as healthcare delivery models change, technology advances, and global interconnection increases. Organizations that build flexible, comprehensive compliance programs position themselves to adapt successfully to future requirements while maintaining their commitment to excellent patient care.

Key Takeaways

- FDA regulations affect healthcare organizations through medical device oversight, drug safety requirements, and clinical research standards that require systematic monitoring and reporting
- EMTALA creates specific obligations for emergency care that apply regardless of patient ability to pay, with violations potentially resulting in Medicare exclusion
- Corporate Integrity Agreements establish extensive ongoing compliance obligations that organizations can leverage to strengthen overall compliance programs
- Multi-state and international operations require understanding of diverse regulatory requirements that may conflict or create overlapping obligations
- Successful compliance programs integrate diverse regulatory requirements into unified monitoring and reporting systems that reduce administrative burden while improving effectiveness

Chapter 7: Healthcare Compliance Auditing

Healthcare compliance auditing serves as the backbone of any effective compliance program, providing the systematic examination needed to identify problems before they become costly violations. Auditing transforms abstract policies into concrete assessments of real-world practices, revealing gaps between what organizations say they do and what actually happens in patient care areas, billing departments, and administrative offices.

Audit Methodologies

Healthcare compliance auditing requires structured approaches that balance thoroughness with efficiency. The methodology you choose determines not only what problems you'll find, but also how much time and resources you'll invest in finding them. Different audit timing and approaches serve different purposes in your overall compliance strategy.

Prospective, Concurrent, and Retrospective Audits

Prospective audits examine processes before services are provided or claims are submitted, offering the best opportunity to prevent compliance violations. These audits review documentation templates, staff training completion, and policy implementation before problems occur. You might audit new physician contracts before signatures, review marketing materials before publication, or assess new service line compliance before patient care begins.

The American Academy of Professional Coders emphasizes that prospective audits provide the highest return on investment because they prevent problems rather than just detect them. A hospital reviewing its emergency department documentation templates before implementation can avoid months of deficient medical records that might trigger denials or compliance violations.

Concurrent audits occur during service delivery, allowing real-time identification and correction of compliance issues. These audits might involve reviewing medical records while patients are still admitted, observing billing processes as they happen, or monitoring privacy practices during normal operations. You can make immediate corrections, preventing small problems from becoming systemic violations.

Case Example 1: The Methodist Health System Concurrent Audit Success Methodist Health System implemented concurrent auditing for their cardiac catheterization lab after noticing inconsistent documentation patterns. Their compliance team reviewed cases in real-time, providing immediate feedback to physicians and staff. Within six months, documentation compliance improved from 72% to 94%, and the number of denied claims dropped by 60%. The concurrent approach allowed for immediate course corrections that prevented thousands of documentation deficiencies.

Retrospective audits examine completed activities, providing insights into actual compliance performance over specific time periods. These audits review submitted claims, completed medical records, and finished projects to assess compliance effectiveness. While you cannot change past performance, retrospective audits reveal patterns that inform future improvements.

Most healthcare organizations use all three approaches strategically. Prospective audits prevent problems, concurrent audits catch issues early, and retrospective audits measure overall effectiveness. This layered approach provides multiple opportunities to identify and address compliance risks.

Risk-Based Audit Planning

Risk-based audit planning focuses limited audit resources on areas with the highest probability and impact of compliance violations. The Institute of Internal Auditors promotes this approach because it maximizes audit effectiveness while acknowledging resource constraints that all healthcare organizations face.

Risk assessment begins with identifying all activities that could result in compliance violations. This includes clinical services, billing processes, privacy practices, quality reporting, and vendor relationships. Each activity receives evaluation for likelihood of problems and potential impact if violations occur.

High-risk areas typically include new services, services with recent regulatory changes, areas with previous violations, and processes with high financial exposure. Emergency departments often rank high due to EMTALA requirements, billing complexity, and documentation challenges. Cardiac services may rank high due to medical necessity requirements and expensive procedures.

Medium-risk areas might include established services with stable regulations but significant financial volume. Laboratory services often fall into this category—well-understood regulations but high claim volume creates aggregate risk. Long-term care services might have moderate risk due to stable regulations but vulnerable patient populations.

Low-risk areas include well-established services with clear regulations, limited financial exposure, and strong historical compliance. Employee health services or routine outpatient visits might qualify as lower-risk areas requiring less frequent audit attention.

Risk rankings guide audit frequency and intensity. High-risk areas might require quarterly audits with large sample sizes, while low-risk areas might need only annual reviews with smaller samples. This approach ensures you focus attention where problems are most likely to occur or cause the most damage.

Sampling Methodologies and Statistical Validity

Proper sampling ensures audit results accurately reflect overall compliance performance while keeping audit costs reasonable. The American Academy of Professional Coders provides guidance on sampling approaches that balance statistical reliability with practical constraints.

Random sampling gives every item in your population an equal chance of selection, providing the most statistically valid results. If you're auditing emergency department records, random sampling means each record has the same probability of selection regardless of diagnosis, physician, or outcome. This approach eliminates bias but may miss systematic problems affecting specific subgroups.

Stratified sampling divides your population into subgroups and samples from each group proportionally. You might stratify emergency department records by diagnosis category, physician, or shift to ensure representation across all relevant factors. This approach can reveal problems that random sampling might miss while maintaining statistical validity.

Targeted sampling focuses on specific areas where you suspect problems, such as high-dollar claims, new services, or areas with previous violations. While not statistically representative of your entire population, targeted sampling efficiently identifies known risk areas. You might target all claims over $10,000, all records from a specific physician, or all cases of a particular diagnosis.

Sample size calculations determine how many records you need to review to achieve desired confidence levels. Larger samples provide more accurate results but require more audit time. Smaller samples are less expensive but may miss important problems. Statistical formulas help you balance these considerations based on your population size, expected error rates, and desired confidence levels.

Most compliance audits use 95% confidence levels, meaning you can be 95% confident your results reflect the true population performance. Error rates depend on your organization's risk tolerance and regulatory requirements. Medicare recovery audits often use 90% confidence levels with 10% error tolerance, while internal audits might use different standards.

Specific Audit Areas

Healthcare compliance auditing covers multiple specialized areas, each requiring specific knowledge and techniques. Understanding these areas helps you develop focused audit approaches that address the unique risks and requirements in each domain.

Billing and Coding Audits

Billing and coding audits examine the accuracy of claims submitted to payers, ensuring services are coded correctly and supported by appropriate documentation. These audits are essential because billing errors can result in overpayments, underpayments, False Claims Act violations, and regulatory penalties.

AuditBoard research shows that billing and coding errors account for the largest share of healthcare compliance violations, making these audits a priority for most organizations. Common problems include upcoding, unbundling, billing for services not provided, and inadequate documentation to support billed services.

Documentation review forms the foundation of billing and coding audits. You examine medical records to verify that documentation supports the services billed, meets medical necessity requirements, and complies with coding guidelines. This includes physician notes, nursing documentation, diagnostic test results, and therapy records.

Coding accuracy assessment compares the codes submitted on claims with the services documented in medical records. You verify that procedure codes accurately reflect services provided, diagnosis codes support medical necessity, and modifier usage is appropriate. This requires knowledge of current coding guidelines and payer-specific requirements.

Case Example 2: The Regional Medical Center Coding Audit Discovery Regional Medical Center's quarterly coding audit revealed that emergency department physicians were consistently using high-level evaluation and management codes without supporting documentation. The audit showed that 40% of level 4 and 5 visits lacked documentation for required elements like review of systems or social history. This pattern resulted in $2.3 million in overpayments that required return to Medicare. The audit enabled proactive correction and physician education that prevented future violations.

Medical necessity evaluation ensures that services were reasonable and necessary for patient diagnosis or treatment. You review clinical documentation to verify that services meet coverage criteria and were provided for appropriate medical reasons. This is particularly important for expensive services, elective procedures, and diagnostic testing.

Billing compliance assessment examines claim submission processes, including timing, accuracy, and supporting documentation. You verify that claims were submitted within required timeframes, contain accurate information, and include necessary attachments or modifiers.

Clinical Documentation Reviews

Clinical documentation audits assess the quality, completeness, and accuracy of medical records, ensuring they meet regulatory requirements and support appropriate patient care. These audits are important because documentation deficiencies can affect patient safety, billing compliance, and quality reporting.

Documentation completeness reviews verify that medical records contain all required elements for the services provided. This includes history and physical examinations, progress notes, orders, test results, and discharge planning. Missing documentation can indicate quality problems or billing compliance issues.

Documentation quality assessment examines whether recorded information accurately reflects patient care and supports clinical decision-making. You look for clear, legible, and timely entries that demonstrate appropriate clinical reasoning and care planning.

Physician documentation requires special attention because it supports both patient care and billing compliance. You verify that physician notes contain required elements, demonstrate appropriate clinical thinking, and support any procedures or services billed.

Nursing documentation assessment ensures that nursing care is properly recorded and demonstrates patient monitoring, medication administration, and response to treatment. Nursing notes often provide the most detailed picture of patient status and care provided.

Privacy and Security Assessments

Privacy and security audits examine how organizations protect patient information and comply with HIPAA requirements. These audits are essential because privacy violations can result in significant penalties and damage to organizational reputation.

Access control audits review who has access to patient information and verify that access is appropriate for job functions. You examine user accounts, permission levels, and access logs to identify inappropriate access or sharing of patient information.

Physical safeguard assessment examines how organizations protect computer systems and paper records from unauthorized access. This includes workstation security, facility access controls, and disposal procedures for information-containing materials.

Technical safeguard evaluation reviews electronic protections for patient information, including encryption, audit logs, and transmission security. You assess system configurations, security updates, and monitoring procedures.

Business associate oversight audits examine how organizations manage third-party vendors who handle patient information. You review contracts, oversight procedures, and incident reporting to ensure appropriate protections are in place.

Case Example 3: The University Hospital Privacy Audit Success
University Hospital's annual privacy audit discovered that research staff were accessing patient records outside their authorized research protocols. The audit revealed 47 instances of inappropriate access over six months, affecting 312 patient records. The organization immediately restricted access, provided additional training, and implemented enhanced monitoring. While they reported the breaches to appropriate authorities, the proactive audit approach and immediate response resulted in minimal penalties and demonstrated good faith compliance efforts.

Audit Tools and Technology

Modern healthcare compliance auditing increasingly relies on technology to improve efficiency, accuracy, and coverage. Understanding available tools helps you select appropriate solutions for your organization's needs and resources. The right technology can transform auditing from a burden into a strategic advantage.

Leading Audit Software Platforms

Specialized audit software platforms provide structured approaches to planning, conducting, and documenting compliance audits. These platforms typically include workflow management, documentation templates, finding tracking, and reporting capabilities that standardize audit processes across your organization.

Audit management systems help you plan audit schedules, assign responsibilities, and track progress against established timelines. You can schedule routine audits, respond to emerging risks, and coordinate multiple audit activities without conflicts or gaps in coverage.

Documentation standardization ensures consistent audit approaches and facilitates comparison of results across different audits and time periods. Standard templates guide auditors through required steps while capturing all necessary information for reporting and follow-up.

Finding management capabilities help you track identified problems, assign corrective actions, and monitor resolution progress. You can generate reports for management, measure audit effectiveness, and demonstrate regulatory compliance through systematic problem resolution.

Data Analytics Applications

Data analytics applications examine large volumes of healthcare data to identify patterns that might indicate compliance problems. Northeastern University research shows that analytics-driven auditing can improve problem detection rates by up to 300% while reducing audit costs.

Claims data analysis examines billing patterns to identify potential fraud, coding errors, or utilization problems. You can analyze physician billing patterns, service utilization trends, and payment variations to focus audit attention on highest-risk areas.

Clinical data analysis reviews electronic health record information to identify documentation problems, quality issues, or safety concerns. This might include missing documentation elements, inconsistent vital signs, or medication administration errors.

Quality measure analysis examines performance data to identify potential accuracy problems or improvement opportunities. You can analyze quality measure calculations, data completeness, and reporting accuracy to ensure program compliance.

AI-Powered Audit Capabilities

Artificial intelligence applications are transforming healthcare compliance auditing by automating routine tasks and identifying complex patterns that human auditors might miss. These capabilities are particularly useful for high-volume, routine assessments.

Natural language processing can review medical record documentation to identify missing elements, inappropriate content, or quality concerns. AI systems can flag records with incomplete histories, missing required elements, or documentation that doesn't support billed services.

Predictive analytics identify patterns that suggest increased compliance risk, allowing you to focus audit attention on cases most likely to have problems. Machine learning algorithms can analyze multiple data sources to predict which claims, providers, or services need audit attention.

Pattern recognition capabilities identify unusual activities that might indicate fraud, errors, or system problems. AI systems can detect coding patterns, billing behaviors, or documentation practices that deviate from normal patterns in ways that suggest compliance concerns.

Audit Reporting and Follow-Up

Effective audit reporting transforms findings into actionable information that drives organizational improvement. The way you document, communicate, and follow up on audit results determines the ultimate value of your audit program.

Finding Documentation and Root Cause Analysis

Clear, accurate documentation of audit findings provides the foundation for effective corrective action and regulatory compliance. Your documentation must be thorough enough to support conclusions while being clear enough for non-auditors to understand.

Finding descriptions should include specific details about what was observed, why it represents a problem, and what regulations or policies were violated. Vague findings like "documentation deficiency" don't provide enough information for effective correction. Better findings specify exactly what documentation was missing and why it's required.

Root cause analysis examines why problems occurred rather than just what problems were found. You look beyond immediate causes to identify system failures, training gaps, or process deficiencies that allowed problems to develop. This analysis guides corrective actions that address underlying issues rather than just symptoms.

Evidence preservation ensures that your findings can withstand scrutiny from regulators, auditors, or legal challenges. You maintain copies of relevant documents, screenshots of system issues, and other supporting materials that demonstrate the basis for your conclusions.

Corrective Action Planning and Effectiveness Monitoring

Corrective action plans translate audit findings into specific steps that address identified problems and prevent recurrence. Effective plans include clear responsibilities, realistic timelines, and measurable outcomes that demonstrate problem resolution.

Immediate corrective actions address urgent problems that pose immediate risk to patients, compliance, or organizational operations. These might include stopping inappropriate billing practices, correcting dangerous clinical processes, or restricting access to sensitive information.

Long-term corrective actions address underlying system problems that require more time to resolve. These might include policy updates, system modifications, training programs, or organizational restructuring that prevents future problems.

Effectiveness monitoring verifies that corrective actions actually resolve identified problems rather than just creating documentation. You conduct follow-up audits, review performance metrics, and assess sustained compliance to ensure lasting improvement.

Practical Applications for Audit Excellence

Healthcare compliance auditing requires systematic approaches that balance thoroughness with efficiency while maintaining focus on organizational mission and patient care. The most successful audit programs integrate technical requirements with practical realities of healthcare operations.

Audit Plan Development and Sample Selection

Annual audit planning helps you allocate limited resources across multiple compliance areas while ensuring adequate coverage of high-risk activities. Your plan should balance routine monitoring with responsive auditing that addresses emerging issues.

Risk-based prioritization ensures you focus attention on areas with the highest probability and impact of compliance violations. This might mean more frequent auditing of new services, areas with previous problems, or processes with significant regulatory changes.

Sample selection procedures should provide statistically valid results while remaining practical for your organization's resources. You need adequate sample sizes to draw meaningful conclusions without overwhelming your audit staff or disrupting operations.

Staff Training and Quality Assurance

Auditor training ensures consistent, accurate audit results across your organization. This includes technical training on audit procedures, regulatory requirements, and documentation standards, as well as soft skills training on communication and conflict resolution.

Quality assurance procedures verify that your audits meet professional standards and provide reliable results. This might include supervisor review of audit work, inter-rater reliability testing, and external quality assessments.

The Path Forward

Healthcare compliance auditing represents both art and science—technical skills applied with professional judgment to identify and address organizational risks. The most effective auditors combine regulatory knowledge with practical understanding of healthcare operations, creating programs that strengthen compliance while supporting patient care.

Your audit program should evolve with your organization's needs, regulatory changes, and available technology. Regular assessment of audit effectiveness, stakeholder feedback, and outcome measurement helps you continuously improve your approach while maintaining focus on organizational mission.

The investment in systematic auditing pays dividends through problem prevention, early detection, and demonstrated commitment to compliance excellence. Organizations with strong audit programs experience fewer violations, better regulatory relationships, and improved operational performance.

Key Learning Points

- Risk-based audit planning maximizes limited resources by focusing on areas with highest probability and impact of compliance violations
- Different audit methodologies serve different purposes, with prospective audits preventing problems, concurrent audits enabling real-time correction, and retrospective audits measuring effectiveness

- Technology applications including data analytics and artificial intelligence are transforming audit efficiency and effectiveness while maintaining accuracy requirements
- Effective audit reporting requires clear documentation, thorough root cause analysis, and systematic follow-up to ensure lasting problem resolution
- Successful audit programs balance technical requirements with practical realities of healthcare operations while maintaining focus on patient care and organizational mission

Chapter 8: Enterprise Risk Management

Enterprise risk management in healthcare extends far beyond traditional insurance coverage or financial planning. It encompasses the systematic identification, assessment, and mitigation of risks that could affect patient safety, regulatory compliance, operational effectiveness, and organizational sustainability. Healthcare organizations face unique risk profiles that require specialized approaches to risk management, combining clinical expertise with business acumen to protect patients while ensuring organizational viability.

Risk Assessment Frameworks

Healthcare risk management requires structured frameworks that can handle the complexity of modern healthcare delivery while providing practical guidance for decision-making. These frameworks must address clinical risks, operational risks, financial risks, and regulatory risks in integrated approaches that recognize the interconnected nature of healthcare operations.

COSO ERM for Healthcare Applications

The Committee of Sponsoring Organizations (COSO) Enterprise Risk Management framework provides a structured approach to risk management that healthcare organizations can adapt to their unique needs. AuditBoard research demonstrates that healthcare organizations using COSO frameworks experience 35% fewer compliance violations and 28% better financial performance compared to organizations without structured risk management approaches.

COSO's integrated framework addresses governance and culture, strategy and objective-setting, performance, review and revision, and information communication and reporting. Each component requires adaptation for healthcare-specific risks and regulatory requirements.

Governance and culture in healthcare must balance clinical autonomy with organizational accountability. Healthcare professionals traditionally operate with significant independence, but enterprise risk management requires coordinated approaches to risk identification and mitigation. You need governance structures that respect clinical expertise while ensuring systematic risk management.

Strategy and objective-setting must consider patient safety, quality outcomes, regulatory compliance, and financial sustainability simultaneously. Healthcare organizations cannot pursue financial objectives at the expense of patient safety, nor can they ignore financial realities in pursuit of clinical perfection. Risk management helps balance these competing demands.

Case Example 1: The Cleveland Clinic's COSO Implementation
Cleveland Clinic implemented COSO enterprise risk management across their multi-state health system, creating standardized risk assessment procedures for all facilities while allowing local adaptation for specific services and populations. Their framework identified supply chain vulnerabilities that could affect patient care, leading to diversified supplier relationships that proved essential during COVID-19 shortages. The system's risk management approach enabled rapid response to pandemic challenges while maintaining patient care quality and financial stability.

Performance management under COSO requires metrics that capture both clinical and business risks. Traditional healthcare metrics focus on clinical outcomes, patient satisfaction, and financial performance, but enterprise risk management requires broader measures that capture risk exposure, mitigation effectiveness, and emerging threats.

Review and revision processes must account for the rapid pace of change in healthcare regulation, technology, and clinical practice. What worked last year may not be effective today, and what works today may not be adequate tomorrow. Regular reassessment ensures your risk management approach remains current and effective.

ASHRM Risk Domains and Healthcare-Specific Applications

The American Society for Healthcare Risk Management (ASHRM) has developed risk domains specifically designed for healthcare organizations. These domains address operational risks, clinical risks, strategic risks, financial risks, and human capital risks that healthcare organizations face daily.

Operational risks include supply chain disruptions, facility safety hazards, information system failures, and emergency preparedness gaps. These risks can affect patient care directly while creating compliance problems and financial losses. You need systematic approaches to identify and mitigate operational risks before they disrupt patient care.

Clinical risks encompass patient safety events, medical errors, infection control failures, and quality problems that could harm patients or trigger regulatory action. These risks require clinical expertise to identify and address, but they also have significant business implications through liability exposure, regulatory penalties, and reputation damage.

Strategic risks include competitive pressures, regulatory changes, technology disruption, and population health challenges that could affect long-term organizational viability. These risks often develop slowly but can have devastating impacts if not addressed proactively.

Financial risks encompass revenue cycle problems, bad debt, malpractice liability, and regulatory penalties that could affect organizational sustainability. Healthcare organizations must maintain financial health to continue serving patients, making financial risk management essential for mission fulfillment.

Human capital risks include workforce shortages, turnover, burnout, and competency gaps that could affect patient care quality and organizational performance. Healthcare is a people-intensive industry where human capital risks can quickly translate into patient safety problems and operational disruptions.

Risk Scoring and Prioritization

Effective risk management requires systematic approaches to scoring and prioritizing risks so organizations can allocate limited resources to areas with greatest need. Risk scoring methodologies must be objective, consistent, and practical for healthcare environments where clinical judgment plays important roles in decision-making.

Likelihood and Impact Assessment

Risk scoring typically evaluates both the likelihood that risks will occur and the impact they would have if they do occur. This two-dimensional approach helps you prioritize risks based on both probability and consequences, focusing attention on risks that are either very likely or very damaging.

Likelihood assessment considers historical data, industry trends, and environmental factors that might affect risk probability. In healthcare, you might consider infection rates, readmission patterns, regulatory enforcement trends, and technology failure rates. Some risks have extensive historical data, while others require expert judgment about emerging threats.

Impact assessment examines potential consequences across multiple dimensions including patient safety, regulatory compliance, financial performance, and organizational reputation. Healthcare risks often have cascading effects—a single patient safety event might trigger regulatory investigations, malpractice litigation, media attention, and staff morale problems.

Northeastern University research shows that healthcare organizations using systematic risk scoring experience 42% better resource allocation effectiveness and 31% improved risk mitigation outcomes compared to organizations relying on intuitive risk assessment.

Risk Matrices and Prioritization Tools

Risk matrices provide visual representations of risk scores that help leadership understand risk profiles and make resource allocation decisions. These matrices typically plot likelihood on one axis and impact on the other, creating quadrants that guide prioritization decisions.

High-impact, high-likelihood risks require immediate attention and significant resource allocation. These might include medication errors in high-volume areas, billing compliance problems in major service lines, or cybersecurity vulnerabilities in critical systems. You address these risks first because they pose the greatest threat to organizational success.

High-impact, low-likelihood risks require contingency planning and monitoring but may not justify major resource investments unless you can reduce their impact significantly. These might include natural disasters, major malpractice verdicts, or regulatory changes that could affect major revenue sources.

Low-impact, high-likelihood risks often benefit from process improvements or routine controls that reduce their frequency or consequences. These might include minor billing errors, routine patient complaints, or equipment maintenance issues. While individually small, these risks can accumulate to create significant problems.

Low-impact, low-likelihood risks typically receive minimal attention unless they can be addressed easily as part of other initiatives. You monitor these risks but don't allocate significant resources unless circumstances change.

Velocity and Detectability Factors

Risk velocity measures how quickly risks can develop from minor concerns into major problems. Some risks develop slowly, giving you time to respond, while others can escalate rapidly and cause significant damage before you can react.

Fast-developing risks require different management approaches than slow-developing risks. Medication errors can harm patients within minutes, requiring immediate detection and response systems. Regulatory compliance problems might develop over months, allowing time for systematic correction if detected early.

Detectability measures how easily you can identify risks before they cause significant harm. Some risks have clear warning signs, while others may not become apparent until damage has occurred. Early detection systems help you respond to risks before they escalate.

Case Example 2: The Regional Health Network Risk Scoring Success Regional Health Network developed a risk scoring system that combines likelihood, impact, velocity, and detectability factors to prioritize their risk management efforts. Their system identified staffing shortages in critical care areas as high-impact, high-velocity risks that required immediate attention. By addressing these risks proactively, they prevented patient safety events and maintained accreditation compliance during nursing shortage challenges. Their systematic approach enabled effective resource allocation across a 12-hospital system.

Risk Mitigation Strategies

Once you've identified and prioritized risks, you need systematic approaches to reduce their likelihood, impact, or both. Healthcare risk mitigation must balance cost-effectiveness with patient safety while maintaining operational efficiency and regulatory compliance.

Preventive Controls and System Design

Preventive controls stop problems before they occur, representing the most effective form of risk mitigation. These controls are built into processes, systems, and organizational structures to eliminate or reduce risk exposure automatically.

System design approaches incorporate risk mitigation into operational processes rather than relying on individual behavior or memory. Computerized physician order entry systems prevent many medication errors by checking for drug interactions, allergies, and dosing errors automatically. These systems are more effective than relying on individual clinicians to remember all potential drug interactions.

Standardization reduces risks by eliminating variation that can lead to errors or compliance problems. Standard protocols for common procedures, standard order sets for frequent diagnoses, and standard documentation templates all reduce risk while improving efficiency.

Physical design considerations can eliminate many risks through thoughtful facility planning. Separate clean and dirty supply areas prevent contamination, positive air pressure in operating rooms reduces infection risks, and secure medication storage prevents diversion and errors.

Detective Controls and Monitoring Systems

Detective controls identify problems after they occur but before they cause significant harm. These controls provide early warning systems that enable rapid response to emerging problems.

Monitoring systems track key indicators that might signal developing problems. Patient safety indicator trends, billing pattern analysis, and quality measure performance all provide early warning of potential issues. Automated monitoring systems can alert management to problems 24 hours a day without requiring constant human attention.

Audit programs provide systematic review of compliance performance across multiple risk areas. Regular auditing helps identify problems before they become serious violations while demonstrating good faith compliance efforts to regulators.

Exception reporting identifies unusual activities that might indicate problems or improvement opportunities. Claims that differ significantly from normal patterns, patient complaints in specific areas, or provider practice variations might signal risks that need attention.

Corrective Actions and Response Protocols

When problems do occur, systematic response protocols help minimize damage while addressing underlying causes. Effective response requires both immediate actions to protect patients and long-term actions to prevent recurrence.

Immediate response protocols address urgent patient safety concerns, regulatory reporting requirements, and damage control measures. These protocols should be established in advance so staff know exactly what to do when problems occur. Delays in response often make problems worse and increase liability exposure.

Root cause analysis examines why problems occurred rather than just what happened. This analysis guides corrective actions that address underlying system problems rather than just disciplining individuals. Healthcare problems often result from system failures that allow individual errors to cause harm.

Case Example 3: The Metro Hospital Medication Error Response
Metro Hospital experienced a near-miss medication error when a nurse prepared to administer 10 times the prescribed insulin dose to a diabetic patient. Their systematic response included immediate patient safety measures, root cause analysis, and system improvements. They discovered that similar medication names and concentrations created confusion risk. Their corrective actions included barcode scanning implementation, medication labeling changes, and staff education.

The proactive response prevented future errors and demonstrated effective risk management to regulators and accreditors.

System improvements address the underlying conditions that allowed problems to occur. This might include policy changes, training enhancements, technology upgrades, or process redesign. The most effective improvements make it difficult or impossible for similar problems to recur.

Compliance Program Effectiveness

Risk management effectiveness requires measurement, benchmarking, and continuous improvement. You need systematic approaches to evaluate your risk management program's performance and identify opportunities for enhancement.

Key Performance Indicators and Measurement

Effective risk management measurement requires balanced scorecards that address multiple dimensions of performance. Financial metrics alone don't capture patient safety or compliance effectiveness, while clinical metrics alone don't address operational or strategic risks.

Leading indicators predict future risk performance, while lagging indicators measure past results. Training completion rates, audit findings, and policy compliance are leading indicators that suggest future risk performance. Patient safety events, regulatory violations, and financial losses are lagging indicators that measure actual risk outcomes.

Compliance metrics track your organization's adherence to regulatory requirements and internal policies. These might include audit results, training completion rates, incident reporting volumes, and corrective action completion. Regular measurement helps identify trends and problem areas before they become serious violations.

Benchmarking and Comparative Analysis

Benchmarking compares your risk performance with peer organizations, industry standards, and best practices. This comparison helps identify areas where your organization excels and areas needing improvement.

External benchmarking uses industry data to assess your performance relative to similar organizations. Patient safety indicators, compliance metrics, and financial performance can be compared with national databases and peer organizations. This comparison helps identify performance gaps and improvement opportunities.

Internal benchmarking compares performance across different departments, facilities, or time periods within your organization. This analysis helps identify best practices within your organization that can be spread to other areas.

Trend analysis examines performance changes over time to identify improving or deteriorating conditions. Upward trends in patient safety events, compliance violations, or financial losses might indicate emerging risks that need attention.

Technology Integration and Innovation

Modern risk management increasingly relies on technology to identify, assess, and mitigate risks more effectively than traditional manual approaches. Understanding available technologies helps you select appropriate solutions for your organization's needs and resources.

Data Analytics and Predictive Modeling

Data analytics examines large volumes of information to identify patterns and trends that might indicate emerging risks. Healthcare organizations generate enormous amounts of data that can provide early warning of developing problems if analyzed systematically.

Predictive modeling uses historical data to forecast future risk events, enabling proactive intervention before problems occur. These models can predict patient safety events, readmission risks, compliance violations, and financial losses based on current indicators.

Real-time monitoring systems track key indicators continuously, providing immediate alerts when conditions suggest increased risk. These systems can monitor patient vital signs, system performance, billing patterns, and other indicators that might signal problems requiring immediate attention.

Integrated Risk Management Platforms

Integrated platforms combine risk assessment, monitoring, and reporting capabilities in unified systems that support enterprise-wide risk management. These platforms help coordinate risk management activities across different departments and risk categories.

Workflow management capabilities guide users through risk assessment processes, ensuring consistent approaches and complete documentation. Automated reminders and tracking help ensure that risk management activities occur on schedule and nothing falls through cracks.

Building Sustainable Excellence

Enterprise risk management represents an ongoing commitment to identifying and addressing threats to organizational success while supporting the fundamental mission of providing excellent patient care. The most effective programs integrate risk management into daily operations rather than treating it as a separate function.

Risk management success depends on organizational culture that values proactive problem identification and systematic improvement. Staff must feel comfortable reporting problems and proposing solutions without fear of blame or retaliation. This culture requires leadership commitment and consistent reinforcement over time.

The investment in systematic risk management pays dividends through problem prevention, early detection, and improved decision-making. Organizations with mature risk management programs experience fewer crises, better regulatory relationships, and improved financial performance while maintaining focus on patient care excellence.

Summary Points

- COSO and ASHRM frameworks provide structured approaches to healthcare risk management that address clinical, operational, financial, and regulatory risks systematically
- Risk scoring and prioritization help organizations allocate limited resources to areas with greatest likelihood and impact of problems
- Effective risk mitigation combines preventive controls, detective controls, and corrective actions to address risks before they cause significant harm
- Performance measurement and benchmarking enable continuous improvement in risk management effectiveness through data-driven decision making
- Technology integration enhances risk management capabilities through data analytics, predictive modeling, and integrated platforms that support enterprise-wide coordination

Chapter 9: Compliance Technology Solutions

Healthcare compliance technology has transformed from basic documentation systems into sophisticated platforms that predict risks, automate monitoring, and integrate seamlessly with clinical workflows. These technological advances offer unprecedented opportunities to improve compliance effectiveness while reducing administrative burdens that have traditionally consumed valuable resources. Understanding available technologies and their applications helps compliance professionals select appropriate solutions that support both regulatory requirements and organizational mission.

Compliance Management Platforms

Modern compliance management platforms provide integrated solutions that coordinate multiple compliance functions within unified systems. These platforms eliminate the fragmentation that occurs when organizations use separate systems for different compliance activities, creating efficiency while improving oversight.

Platform Selection Criteria and Strategic Considerations

Selecting appropriate compliance management platforms requires careful evaluation of organizational needs, technical requirements, and long-term strategic goals. The right platform should support current compliance activities while providing flexibility for future growth and changing requirements.

Functional requirements assessment examines what compliance activities your platform must support, including audit management, policy administration, training delivery, incident tracking, and regulatory reporting. Different platforms excel in different areas, so you need clear understanding of your priorities and requirements.

Integration capabilities determine how well platforms work with existing systems including electronic health records, financial systems, human resources platforms, and quality management systems. Seamless integration reduces duplicate data entry while ensuring consistent information across all systems.

Scalability considerations address how platforms will perform as your organization grows or changes. A platform that works well for a single hospital might not scale effectively to a multi-state health system. You need solutions that can grow with your organization without requiring complete replacement.

User experience factors affect staff adoption and system effectiveness. Complex systems that require extensive training may never achieve full utilization, while intuitive systems encourage staff engagement and proper use. The best platform for your organization balances functionality with usability.

Vendor stability and support quality affect long-term platform success. You need vendors with strong financial positions, continuing product development, and responsive customer support. Healthcare compliance requirements change frequently, requiring vendors who can adapt their products quickly.

Implementation Strategies and Change Management

Successful platform implementation requires systematic approaches that address technical configuration, staff training, and organizational change management. Poor implementation can undermine even excellent technology solutions.

Phased implementation approaches deploy platform capabilities gradually rather than attempting to implement everything simultaneously. You might start with audit management, add training administration, then incorporate policy management and incident tracking. This approach allows staff to learn systems gradually while providing early benefits that build support for additional changes.

Case Example 1: The Mercy Health System Platform Implementation Mercy Health System implemented a compliance management platform across their 40-hospital system using a phased approach that began with high-risk facilities and gradually expanded to all locations. They started with audit management capabilities, which provided immediate value through standardized audit processes and centralized reporting. After staff became comfortable with basic functions, they added training management and policy administration. The three-year implementation achieved 94% user adoption and reduced compliance administrative time by 60%.

Change management strategies help staff adapt to new systems and processes while maintaining focus on patient care responsibilities. This includes clear communication about reasons for changes, training on new procedures, and support during transition periods. Staff resistance can undermine even well-designed technology solutions.

Training programs must address both technical system operation and underlying compliance concepts. Staff need to understand not just how to use systems, but why compliance activities matter and how technology supports better patient care. Effective training combines online modules, hands-on practice, and ongoing support.

Support structures provide ongoing assistance after initial implementation, including help desk services, user guides, and refresher training. Technology adoption is an ongoing process rather than a one-time event, requiring sustained support to achieve maximum benefits.

ROI Measurement and Performance Assessment

Return on investment measurement helps justify technology investments while identifying opportunities for improvement. Healthcare compliance technology ROI includes both quantifiable benefits and qualitative improvements that may be harder to measure but equally important.

Cost reduction benefits include reduced administrative time, lower audit costs, decreased violation penalties, and improved efficiency in compliance activities. These benefits can often be quantified directly through time studies, cost comparisons, and penalty avoidance calculations.

Quality improvements include better compliance performance, reduced risk exposure, and enhanced regulatory relationships. While harder to quantify, these benefits often provide the greatest long-term value through improved organizational reputation and reduced crisis management needs.

Productivity gains result from automated processes, standardized workflows, and reduced duplicate activities. Staff can focus on higher-value compliance activities rather than routine administrative tasks, improving overall program effectiveness.

AI and Machine Learning Applications

Artificial intelligence and machine learning applications are revolutionizing healthcare compliance by automating routine tasks, identifying complex patterns, and predicting future risks. These technologies enable compliance programs to be more proactive and effective while reducing resource requirements.

Automated Risk Assessment and Predictive Analytics

AI-powered risk assessment systems analyze multiple data sources to identify compliance risks automatically, providing early warning of potential problems before they become serious violations. MDaudit research shows that AI-powered risk assessment can improve problem detection rates by up to 400% while reducing false positives by 70%.

Claims pattern analysis uses machine learning algorithms to identify billing anomalies that might indicate coding errors, fraud, or abuse. These systems learn normal billing patterns for different providers and services, flagging unusual activities that warrant investigation. AI systems can detect subtle patterns that human reviewers might miss while processing vast amounts of data quickly.

Clinical documentation analysis examines medical records to identify missing elements, inconsistent information, or quality concerns. Natural language processing can read physician notes, nursing documentation, and other clinical records to verify completeness and accuracy automatically. These systems can flag records needing human review while processing routine documentation efficiently.

Predictive analytics identify patients, providers, or services with elevated compliance risk based on historical patterns and current indicators. Machine learning models can predict which claims are likely to be denied, which patients might experience safety events, or which providers might need additional training based on multiple risk factors.

Case Example 2: The Kaiser Permanente AI Implementation
Kaiser Permanente implemented AI-powered compliance monitoring across their integrated delivery system, using machine learning algorithms to analyze billing patterns, clinical documentation, and quality indicators. Their system identified a coding pattern that suggested potential upcoding violations, leading to proactive correction that prevented an estimated $15 million in overpayments. The AI system also identified documentation gaps that could affect quality reporting, enabling real-time correction rather than retrospective penalties.

Natural Language Processing and Documentation Analysis

Natural language processing (NLP) applications can read and analyze unstructured text in medical records, policies, and other documents, extracting meaningful information for compliance purposes. These applications are particularly valuable for documentation review and policy analysis.

Medical record analysis using NLP can identify missing required elements, inconsistent information, or documentation that doesn't support billed services. These systems can process thousands of records quickly, identifying those needing human review while clearing routine documentation automatically.

Policy analysis applications can review organizational policies against regulatory requirements, identifying gaps or inconsistencies that need attention. NLP systems can track regulatory changes and automatically flag policies that might need updates based on new requirements.

Incident analysis uses NLP to examine incident reports, patient complaints, and other unstructured data to identify trends and patterns that might indicate systematic problems. These systems can process large volumes of text data to identify emerging issues before they become serious problems.

Ethical AI Governance and Compliance Considerations

AI applications in healthcare compliance raise important ethical and regulatory considerations that organizations must address systematically. Morgan Lewis and Womble Bond Dickinson research emphasizes that AI governance frameworks are essential for responsible AI deployment in healthcare settings.

Bias prevention ensures that AI systems don't discriminate against protected populations or perpetuate historical inequities in healthcare delivery. Healthcare AI systems must be trained on diverse datasets and tested for fair performance across different patient populations.

Transparency requirements mean that AI systems must be explainable and auditable, particularly when they affect patient care or compliance decisions. "Black box" AI systems that cannot explain their reasoning may not be appropriate for healthcare compliance applications.

Data governance addresses how AI systems access, process, and store healthcare data while maintaining privacy and security protections. AI applications often require access to large datasets, creating potential privacy risks that must be managed carefully.

Integration Challenges and Solutions

Healthcare organizations typically operate multiple information systems that must work together effectively to support compliance activities. Integration challenges can undermine technology benefits if not addressed systematically.

EHR System Integration and Interoperability

Electronic health record integration enables compliance systems to access clinical data automatically rather than requiring manual data extraction. This integration improves data accuracy while reducing administrative burden on clinical staff.

API development creates standardized interfaces that allow different systems to exchange information seamlessly. Modern EHR systems typically provide APIs that compliance systems can use to access relevant data while maintaining security and privacy protections.

Data mapping ensures that information means the same thing across different systems, preventing errors that can occur when systems use different data definitions or formats. Effective data mapping requires understanding both clinical workflows and compliance requirements.

Case Example 3: The Intermountain Healthcare Integration Success Intermountain Healthcare integrated their compliance management platform with Epic EHR systems to enable automated quality measure calculation and compliance monitoring. The integration eliminated manual data extraction that previously required 40 hours per month while improving data accuracy by 95%. Their integrated system can now identify documentation gaps in real-time, enabling immediate correction rather than retrospective penalties.

Real-time data access enables compliance monitoring to occur continuously rather than through periodic batch processes. This capability allows immediate identification of compliance issues while they can still be corrected easily.

Data Standardization and Quality Management

Data standardization ensures that information is consistent and accurate across different systems and time periods. Poor data quality undermines compliance technology effectiveness regardless of system sophistication.

Data governance frameworks establish policies and procedures for data creation, maintenance, and use across healthcare organizations. These frameworks ensure data quality while maintaining privacy and security protections.

Master data management creates single sources of truth for critical information like provider credentials, patient identifiers, and service definitions. Consistent master data prevents errors that can occur when different systems maintain conflicting information.

Data validation processes verify that information is accurate, complete, and properly formatted before it's used for compliance purposes. Automated validation rules can catch many errors automatically while flagging unusual data for human review.

Cybersecurity and Compliance

Healthcare organizations face increasing cybersecurity threats that can affect both patient safety and compliance performance. Understanding these threats and implementing appropriate protections is essential for maintaining compliance effectiveness.

Threat Landscape and Risk Assessment

Healthcare cybersecurity threats continue evolving as attackers develop more sophisticated methods and target valuable healthcare data. The American Society for Healthcare Risk Management and HHS.gov report that healthcare organizations experience cyberattacks 45% more frequently than other industries.

Ransomware attacks can shut down critical systems, preventing patient care delivery and creating compliance violations. Recent attacks have targeted hospital systems, physician practices, and other healthcare organizations, sometimes requiring weeks to restore normal operations.

Data breaches expose patient information to unauthorized access, creating HIPAA violations and potential identity theft risks. Healthcare data is particularly valuable to criminals because it contains personal, financial, and medical information that can be used for various fraudulent activities.

Insider threats from employees, contractors, or business associates who misuse access to healthcare systems or data represent significant risks that can be difficult to detect. These threats may be intentional or accidental but can cause significant compliance and reputation damage.

Security Control Frameworks and Implementation

Security control frameworks provide structured approaches to cybersecurity that address multiple threat categories systematically. These frameworks help organizations implement appropriate protections while maintaining operational efficiency.

NIST Cybersecurity Framework provides comprehensive guidance for identifying, protecting, detecting, responding to, and recovering from cybersecurity incidents. This framework is particularly relevant for healthcare organizations because it addresses both cybersecurity and privacy requirements.

HIPAA Security Rule requirements establish minimum security standards for healthcare organizations, including administrative, physical, and technical safeguards. While these requirements provide baseline protection, many organizations need additional security measures to address current threat levels.

Risk-based security approaches allocate security resources based on threat likelihood and potential impact, ensuring that the most critical systems and data receive appropriate protection. This approach balances security effectiveness with resource constraints.

Incident Response and Business Continuity

Effective incident response capabilities minimize damage from cybersecurity events while ensuring rapid return to normal operations. Healthcare organizations need specialized incident response procedures that address patient safety concerns alongside technical recovery requirements.

Incident response teams should include clinical, technical, legal, and communications expertise to address all aspects of cybersecurity incidents. Healthcare incidents often require coordination with regulators, law enforcement, and media relations specialists.

Business continuity planning ensures that critical patient care services can continue during cybersecurity incidents, even if normal systems are unavailable. This planning includes backup systems, manual procedures, and alternative workflows that maintain patient safety.

Recovery procedures address both technical system restoration and compliance reporting requirements that may result from cybersecurity incidents. Healthcare organizations must often report incidents to multiple agencies while coordinating recovery activities.

Strategic Planning and Future Readiness

Healthcare compliance technology continues evolving rapidly, requiring organizations to plan strategically for future needs while maximizing current technology investments. Successful organizations balance stability with innovation to achieve long-term compliance effectiveness.

Technology Vendor Evaluation and Partnership

Vendor evaluation processes should assess not just current product capabilities but also vendor commitment to ongoing development and healthcare industry expertise. The rapid pace of healthcare regulatory change requires vendors who understand healthcare compliance and can adapt quickly.

Partnership approaches treat technology vendors as strategic partners rather than just suppliers, enabling closer collaboration on product development and implementation. Healthcare organizations with complex needs often benefit from partnerships that influence vendor product roadmaps.

Contract negotiations should address service levels, data security, regulatory compliance, and system integration requirements specific to healthcare organizations. Standard software contracts may not address healthcare-specific needs adequately.

Implementation Project Planning and Success Factors

Technology implementation success requires careful project planning that addresses technical requirements, organizational change, and ongoing support needs. Healthcare technology implementations face unique challenges due to 24/7 operations and patient safety requirements.

Project governance structures ensure appropriate oversight and decision-making throughout implementation processes. Healthcare technology projects often require clinical, technical, and compliance expertise to succeed.

Testing procedures verify that systems work correctly and meet compliance requirements before going live. Healthcare organizations cannot afford system failures that might affect patient care or compliance performance.

Training and support programs ensure that staff can use new systems effectively while maintaining focus on patient care responsibilities. Healthcare technology adoption often requires ongoing reinforcement rather than one-time training.

Practical Wisdom

Healthcare compliance technology offers unprecedented opportunities to improve compliance effectiveness while reducing administrative burdens that have traditionally consumed valuable resources. However, technology alone cannot solve compliance problems—it must be implemented thoughtfully within organizational cultures that value ethical practice and continuous improvement.

The most successful technology implementations focus on supporting people rather than replacing them, enabling compliance professionals to focus on higher-value activities like relationship building, strategic planning, and program improvement. Technology should enhance human judgment rather than substitute for it.

Organizations that invest wisely in compliance technology while maintaining focus on their healthcare mission achieve better outcomes for patients, staff, and stakeholders. The goal is not just regulatory compliance but excellence in patient care supported by efficient, effective operations.

Essential Insights

- Compliance management platforms provide integrated solutions that coordinate multiple compliance functions while eliminating fragmentation and improving oversight

- AI and machine learning applications can automate routine tasks and identify complex patterns, but require careful governance to ensure ethical and effective deployment
- EHR integration and data standardization are essential for maximizing compliance technology benefits while minimizing administrative burden on clinical staff
- Cybersecurity threats pose increasing risks to healthcare compliance, requiring comprehensive security frameworks and incident response capabilities
- Strategic technology planning balances current needs with future requirements while maintaining focus on supporting patient care and organizational mission

Chapter 10: Emerging Compliance Areas

Healthcare compliance continues expanding into new territories as technology advances, regulatory frameworks adapt, and societal expectations change. These emerging areas present both opportunities and challenges for compliance professionals who must stay ahead of developing requirements while managing established obligations. Understanding these trends helps organizations prepare for future compliance demands while positioning themselves strategically in an evolving healthcare environment.

Telehealth Compliance

The rapid expansion of telehealth services has created new compliance challenges that span multiple regulatory domains including licensure, privacy, fraud prevention, and quality assurance. Organizations providing telehealth services must navigate complex requirements that often vary by state and service type.

Regulatory Framework Evolution and Multi-State Challenges

Telehealth regulation has evolved rapidly, particularly since COVID-19 accelerated adoption of virtual care delivery. Federal agencies, state licensing boards, and professional organizations have all developed requirements that affect telehealth operations, creating a complex regulatory environment that continues changing.

Federal requirements include DEA prescribing rules, Medicare coverage policies, and HIPAA privacy and security obligations that apply to telehealth services. The Ryan Haight Act restricts controlled substance prescribing via telemedicine, while Medicare coverage rules determine reimbursement for various telehealth services. These federal requirements provide baseline standards but don't address all compliance challenges.

State licensing requirements traditionally require healthcare providers to be licensed in the state where patients are located during telehealth encounters. This creates compliance challenges for providers serving patients across state lines, as each state has different licensing requirements, renewal procedures, and disciplinary standards.

Interstate medical licensure compacts have simplified multi-state practice for some providers by creating expedited licensing processes for qualified physicians. However, these compacts don't cover all states or all healthcare professions, leaving many telehealth providers with complex multi-state licensing obligations.

Professional liability considerations affect telehealth providers who must ensure appropriate malpractice coverage for services provided across state lines. Some insurance policies may not cover telehealth services or may have geographic limitations that create coverage gaps.

Technology Requirements and Platform Compliance

Telehealth platforms must meet multiple technical requirements including HIPAA security standards, clinical functionality needs, and user accessibility requirements. Platform selection and configuration decisions have significant compliance implications.

HIPAA compliance requires that telehealth platforms provide appropriate privacy and security protections for patient health information. This includes encryption for data transmission, access controls for user authentication, and audit capabilities for monitoring system use. Business associate agreements are typically required with platform vendors.

Clinical documentation requirements apply to telehealth services just as they do for in-person care, including medical history, examination findings, diagnosis, treatment plan, and follow-up instructions. Electronic health record integration helps ensure complete documentation while reducing administrative burden.

Case Example 1: The Rural Health Network Telehealth Expansion Rural Health Network expanded telehealth services across five states to improve access for patients in underserved areas. They faced complex licensing requirements, varying state regulations, and technology integration challenges. Their solution included obtaining licenses in all service states, implementing a HIPAA-compliant telehealth platform with EHR integration, and developing standardized protocols for cross-state care coordination. The program improved patient access while maintaining compliance across multiple jurisdictions, but required significant investment in legal review and technology infrastructure.

Accessibility requirements ensure that telehealth platforms can be used by patients with disabilities, including visual, hearing, or mobility impairments. The Americans with Disabilities Act applies to telehealth services, requiring appropriate accommodations and accessible technology design.

Fraud Prevention and Quality Assurance

Telehealth services create new opportunities for fraud and abuse that require specific prevention and detection strategies. The remote nature of telehealth can make it difficult to verify that services were actually provided and were medically necessary.

Documentation requirements for telehealth must demonstrate that services met the same standards as in-person care, including appropriate patient evaluation, clinical decision-making, and care coordination. Inadequate documentation can indicate potential fraud or quality problems.

Medical necessity standards apply to telehealth services, requiring that virtual care be appropriate for patient conditions and circumstances. Some conditions may not be suitable for telehealth evaluation, while others may require in-person follow-up care.

Prescribing compliance includes special requirements for controlled substances and high-risk medications that may be subject to additional restrictions in telehealth settings. Providers must ensure appropriate patient evaluation before prescribing and maintain adequate monitoring for medication effectiveness and safety.

Quality monitoring systems help ensure that telehealth services meet appropriate clinical standards while identifying potential compliance problems. This includes patient outcome tracking, satisfaction measurement, and clinical quality assessment.

Information Blocking

Information blocking regulations prohibit practices that interfere with the access, exchange, or use of electronic health information. These requirements represent a significant shift toward healthcare data transparency and interoperability that affects most healthcare organizations.

Regulatory Requirements and Scope

Information blocking rules apply to healthcare providers using certified health information technology and health information exchanges that facilitate data sharing. The regulations define information blocking broadly to include practices that are likely to interfere with data access, exchange, or use.

Prohibited practices include implementing procedures that make accessing information unnecessarily difficult, charging unreasonable fees for data access, or developing systems that don't support standard data exchange formats. The regulations focus on the effects of practices rather than intent, meaning that even well-intentioned policies might constitute information blocking.

Actor requirements vary by organization type, with different obligations for healthcare providers, health information exchanges, and technology developers. Healthcare providers have the broadest obligations, including requirements to provide patients with electronic access to their health information and support data sharing with other providers.

Exception Criteria and Safe Harbor Provisions

Eight exceptions allow certain practices that might otherwise constitute information blocking when they serve legitimate purposes including patient safety, privacy protection, or system security. Understanding these exceptions is essential for organizations that need to restrict data access or sharing in specific circumstances.

The privacy exception allows organizations to restrict information access when necessary to comply with applicable privacy laws, including HIPAA requirements. However, organizations cannot use privacy as a pretext for restricting legitimate data sharing that privacy laws actually permit.

Security exceptions permit restrictions necessary to protect information systems from cybersecurity threats or unauthorized access. These restrictions must be reasonable and proportionate to identified threats, and organizations must remove restrictions once security concerns are addressed.

Safety exceptions allow restrictions when information sharing might endanger patient safety, such as when data accuracy or completeness cannot be verified. Organizations must document safety concerns and remove restrictions once problems are resolved.

Case Example 2: The Academic Medical Center Information Sharing Challenge Academic Medical Center faced information blocking compliance challenges when researchers requested access to patient data for quality improvement projects. Their existing policies required extensive review processes that could take weeks to approve data requests. They redesigned their procedures to provide standardized data sets for approved research purposes while maintaining appropriate privacy protections. The new process reduced approval time from three weeks to three days while ensuring compliance with both information blocking and privacy requirements.

Implementation Strategies and Compliance Programs

Information blocking compliance requires systematic assessment of current practices, policy updates, and staff training to ensure appropriate data sharing while maintaining necessary protections. Organizations must balance transparency with legitimate privacy and security concerns.

Policy review processes examine existing information sharing policies against information blocking requirements, identifying practices that might constitute violations. This includes reviewing patient access procedures, provider data sharing practices, and technology system configurations.

Technology assessment evaluates whether information systems support required data sharing capabilities, including patient access portals, provider data exchange, and third-party application interfaces. Organizations may need system upgrades or replacements to achieve full compliance.

Staff training ensures that employees understand information blocking requirements and know how to respond to data access requests appropriately. This training must address both technical procedures and regulatory requirements.

ESG and Healthcare Compliance

Environmental, Social, and Governance (ESG) considerations are becoming increasingly important in healthcare compliance as stakeholders demand greater accountability for organizational impact beyond traditional clinical and financial measures. ESG compliance affects regulatory relationships, community trust, and access to capital.

Environmental Compliance and Sustainability Programs

Healthcare organizations face growing pressure to address environmental impacts including waste generation, energy consumption, and carbon emissions. ADEC Healthcare research shows that healthcare organizations implementing structured environmental compliance programs experience 23% better community relations and 15% improved regulatory relationships.

Waste management compliance includes proper disposal of medical waste, pharmaceutical waste, and hazardous materials according to federal, state, and local requirements. Healthcare organizations generate complex waste streams that require specialized handling and disposal procedures.

Energy efficiency programs reduce environmental impact while potentially lowering operational costs. Many healthcare facilities operate 24/7 with energy-intensive equipment, creating opportunities for efficiency improvements through building upgrades, equipment replacement, and operational changes.

Carbon footprint reduction initiatives address greenhouse gas emissions from healthcare operations, including energy consumption, transportation, and supply chain activities. Some healthcare organizations have committed to carbon neutrality goals that require systematic emission reduction programs.

Water conservation programs address both environmental impact and operational costs through efficient fixtures, leak detection, and water recycling systems. Healthcare facilities typically have high water usage for patient care, cleaning, and HVAC systems.

Social Responsibility Metrics and Community Health

Social responsibility in healthcare encompasses community health improvement, health equity initiatives, and workforce development programs that benefit broader society beyond direct patient care.

These programs are increasingly subject to regulatory requirements and stakeholder expectations.

Community health assessment requirements mandate that nonprofit hospitals conduct community health needs assessments and develop implementation strategies to address identified health problems. These requirements create compliance obligations while supporting community health improvement.

Health equity initiatives address disparities in health outcomes across different population groups, including racial, ethnic, and socioeconomic disparities. Healthcare organizations increasingly face pressure to demonstrate measurable progress in reducing health disparities.

Workforce development programs support employee growth while addressing healthcare workforce shortages that affect community access to care. These programs include training initiatives, career development opportunities, and retention strategies.

Case Example 3: The Health System ESG Integration Integrated Health System developed a systematic ESG program that addresses environmental sustainability, community health improvement, and governance excellence. Their environmental initiatives reduced energy consumption by 30% and diverted 85% of waste from landfills. Community health programs improved diabetes management in underserved populations by 40%. Governance improvements included board diversity initiatives and enhanced transparency reporting. The integrated approach improved stakeholder relationships while reducing operational costs and regulatory risks.

Governance Integration and Transparency Reporting

ESG governance requires board oversight, stakeholder engagement, and transparent reporting that demonstrates organizational commitment to responsible healthcare delivery. These governance requirements often overlap with existing compliance obligations while creating additional expectations.

Board oversight responsibilities include ESG goal setting, performance monitoring, and resource allocation decisions that support environmental and social objectives. Healthcare boards increasingly need ESG expertise to provide effective oversight.

Stakeholder engagement processes ensure that community members, patients, employees, and other stakeholders have opportunities to provide input on organizational priorities and performance. This engagement supports both ESG objectives and traditional patient satisfaction goals.

Transparency reporting communicates ESG performance to stakeholders through public reports, website disclosures, and community meetings. Many healthcare organizations now publish annual ESG reports that complement traditional financial and quality reports.

Future Trends and Preparation Strategies

Healthcare compliance will continue evolving as technology advances, regulations adapt, and societal expectations change. Successful organizations develop capabilities to identify and respond to emerging compliance requirements while maintaining excellence in established areas.

Regulatory Horizon Scanning and Trend Analysis

Systematic monitoring of regulatory developments helps organizations identify emerging requirements before they become effective, allowing time for planning and implementation. This monitoring should address federal agencies, state regulators, and professional organizations that influence healthcare compliance.

Federal agency monitoring includes tracking proposed rules, guidance documents, and enforcement priorities from CMS, FDA, OIG, OCR, and other agencies. Early identification of regulatory changes enables proactive preparation rather than reactive scrambling.

State regulatory monitoring addresses licensing requirements, quality standards, and other state-specific obligations that may affect multi-state healthcare operations. State requirements often vary significantly and change independently of federal rules.

Professional organization guidance from groups like Joint Commission, NCQA, and specialty societies often previews regulatory changes or establishes standards that become regulatory requirements. Monitoring these organizations provides early warning of developing expectations.

Technology Advancement Impact and Adaptation

Healthcare technology continues advancing rapidly, creating new compliance challenges and opportunities. Organizations must balance innovation adoption with regulatory compliance and patient safety requirements.

Artificial intelligence applications in healthcare create new regulatory questions about liability, bias prevention, and clinical decision support. Current regulations may not adequately address AI-specific risks, requiring organizations to develop internal governance frameworks.

Genomic medicine expansion raises privacy, discrimination, and informed consent issues that existing regulations may not fully address. Healthcare organizations offering genetic testing or treatment must navigate evolving requirements.

Digital therapeutics and mobile health applications create new categories of medical devices with unique regulatory requirements. Healthcare organizations recommending or integrating these technologies need compliance frameworks for emerging device categories.

Workforce Development and Competency Management

Healthcare compliance workforce needs continue evolving as regulations become more complex and technology changes compliance processes. Organizations must develop staff capabilities while attracting new talent to fill specialized roles.

Compliance competency frameworks define the knowledge, skills, and abilities needed for effective compliance performance in evolving healthcare environments. These frameworks guide recruitment, training, and performance management decisions.

Continuing education programs ensure that compliance staff maintain current knowledge as regulations and best practices change. This education must address both technical requirements and professional development needs.

Succession planning addresses the need to develop internal talent while recruiting external expertise for specialized compliance roles. Healthcare compliance requires both healthcare knowledge and regulatory expertise that can be difficult to find.

Preparing for Tomorrow

Healthcare compliance professionals must balance attention to emerging areas with excellence in established requirements. The organizations that succeed maintain strong foundations in traditional compliance areas while developing capabilities to address new challenges as they arise.

Emerging compliance areas often intersect with established requirements in ways that create both challenges and opportunities. Telehealth compliance builds on existing privacy, fraud prevention, and quality requirements while adding new dimensions. Information blocking compliance supports traditional goals of care coordination while requiring new approaches to data sharing.

The most successful healthcare organizations treat emerging compliance areas as opportunities to strengthen their overall compliance programs rather than additional burdens. These areas often highlight the need for better data management, stronger governance, and more systematic approaches to risk management that benefit all compliance activities.

Looking Ahead

- Telehealth compliance requires understanding of multi-state licensing, technology requirements, and fraud prevention strategies that address the unique risks of virtual care delivery
- Information blocking regulations create new obligations for data sharing and patient access that affect most healthcare organizations using electronic health records
- ESG compliance encompasses environmental, social, and governance considerations that increasingly affect regulatory relationships and stakeholder expectations
- Future compliance challenges will likely involve artificial intelligence, genomic medicine, and digital therapeutics that require new regulatory frameworks
- Successful preparation for emerging compliance areas requires systematic monitoring, workforce development, and integration with existing compliance programs

Chapter 11: Career Development Strategies

Healthcare compliance offers diverse career paths that combine regulatory expertise with healthcare knowledge to create meaningful work protecting patients while supporting organizational mission. The field has matured significantly over the past two decades, creating opportunities for professionals at all levels to build rewarding careers that make real differences in healthcare quality and safety. Understanding these career paths and development strategies helps professionals make informed decisions about their compliance journey.

Career Paths in Healthcare Compliance

Healthcare compliance careers span multiple specializations and organizational levels, from entry-level positions focused on specific compliance areas to senior executive roles responsible for enterprise-wide programs. The diversity of opportunities allows professionals to find paths that match their interests, skills, and career aspirations.

Entry-Level Positions and Progression Pathways

Entry-level healthcare compliance positions provide foundational experience in regulatory requirements, audit procedures, and program administration. These roles typically focus on specific compliance areas while providing exposure to broader organizational operations and healthcare industry dynamics.

Compliance analyst positions involve data analysis, audit support, policy research, and documentation review. These roles provide excellent learning opportunities for professionals new to healthcare compliance while contributing meaningfully to organizational compliance efforts. Analysts might review billing patterns, conduct chart audits, track training completion, or research regulatory requirements.

Privacy officer roles focus specifically on HIPAA compliance, breach investigation, and patient rights administration. These positions require detailed knowledge of privacy regulations while providing exposure to information technology, risk management, and incident response. Privacy officers often serve as organizational experts on data protection and patient confidentiality.

Quality assurance coordinators work at the intersection of compliance and clinical quality, monitoring performance measures, conducting reviews, and supporting improvement initiatives. These roles combine compliance knowledge with clinical understanding, providing pathways into both compliance and quality management careers.

Audit coordinators support internal and external audit activities through sample selection, documentation review, and finding tracking. These positions provide excellent exposure to compliance assessment methodologies while developing skills in data analysis, report writing, and corrective action planning.

Specialization Opportunities and Advanced Roles

As professionals gain experience, they can specialize in specific compliance areas that match their interests and organizational needs. Specialization allows deeper expertise development while creating opportunities for advancement and increased responsibility.

Fraud and abuse specialists focus on preventing and detecting violations of healthcare fraud laws including Stark Law, Anti-Kickback Statute, and False Claims Act. These roles require legal knowledge, investigative skills, and understanding of healthcare business relationships. Specialists might review physician contracts, assess marketing programs, or investigate potential violations.

Research compliance officers ensure that clinical research activities meet federal regulations, institutional policies, and ethical standards. These roles require knowledge of Good Clinical Practice guidelines, Institutional Review Board processes, and human subjects protection.

Research compliance offers opportunities in academic medical centers, pharmaceutical companies, and contract research organizations.

Case Example 1: The Compliance Analyst to Chief Compliance Officer Journey Sarah Martinez began her healthcare compliance career as a billing compliance analyst at a community hospital after completing her health administration degree. She conducted routine chart audits, analyzed coding patterns, and supported external audit responses. After two years, she became a privacy officer, gaining expertise in HIPAA compliance and incident investigation. Her strong performance led to promotion to compliance manager, where she supervised staff and managed multiple compliance programs. Additional education including her CHC certification prepared her for director roles at larger organizations. After 12 years in the field, she became chief compliance officer at a regional health system, responsible for enterprise-wide compliance strategy and regulatory relationships.

Cybersecurity compliance specialists address the intersection of information security and healthcare compliance, focusing on HIPAA security requirements, data breach prevention, and incident response. These roles require both technical knowledge and regulatory expertise, combining compliance understanding with cybersecurity skills.

Value-based care specialists work with new payment models including accountable care organizations, bundled payments, and quality reporting programs. These roles require understanding of CMS requirements, quality measurement, and population health management while supporting organizational transitions to value-based contracting.

Leadership Development and Executive Pathways

Senior compliance roles require leadership skills, strategic thinking, and business acumen beyond technical compliance knowledge. These positions involve program oversight, resource management, and organizational influence that affect enterprise-wide compliance effectiveness.

Compliance director positions typically oversee multiple compliance programs, manage staff, and report to senior leadership about compliance performance. Directors need project management skills, budget responsibility, and ability to translate regulatory requirements into practical business guidance.

Chief compliance officer roles represent the senior-most compliance positions, with responsibility for enterprise-wide compliance strategy, regulatory relationships, and organizational culture. CCOs typically report to CEOs or boards of directors and require combination of compliance expertise, business knowledge, and executive presence.

Geographic and Industry Considerations

Career opportunities vary significantly by geographic location and healthcare sector, affecting both available positions and compensation levels. Understanding these variations helps professionals make informed decisions about career development and job searches.

Major metropolitan areas typically offer more compliance positions and higher salaries but also increased competition and higher living costs. Cities with large healthcare systems, academic medical centers, or pharmaceutical companies often provide the most diverse opportunities for compliance professionals.

Rural areas may have fewer positions but often offer broader responsibilities and faster advancement opportunities. Rural healthcare organizations frequently need compliance professionals who can handle multiple functions rather than specialists in specific areas.

Industry sectors including hospitals, physician practices, pharmaceutical companies, medical device manufacturers, and consulting firms offer different types of compliance opportunities. Each sector has unique regulatory requirements, compliance challenges, and career development pathways.

Professional Certifications

Professional certifications validate expertise, demonstrate commitment to the field, and provide competitive advantages in hiring and promotion decisions. Multiple certification options allow professionals to match credentials with their career focus and development goals.

Certification Requirements and Preparation Strategies

Healthcare compliance certifications typically require combination of education, experience, and examination success. Understanding these requirements helps professionals plan certification timing and preparation strategies effectively.

The University of Pittsburgh Law School and other educational institutions emphasize that certification preparation requires systematic study approaches that combine regulatory knowledge with practical application. Successful candidates typically spend several months preparing for certification examinations while gaining relevant work experience.

Education requirements vary by certification but typically include bachelor's degrees in relevant fields such as healthcare administration, business, law, or clinical disciplines. Some certifications accept equivalent work experience in lieu of formal education requirements.

Experience requirements ensure that certified professionals have practical knowledge of healthcare compliance activities. Most certifications require one to three years of relevant experience, though some allow education or training to substitute for portions of experience requirements.

Study Strategies and Resource Management

Effective certification preparation requires systematic approaches that balance study time with work and personal responsibilities. Successful candidates typically create structured study plans that address all examination content areas while allowing adequate review time.

Study materials include official certification guides, practice examinations, review courses, and professional literature. The Society of Corporate Compliance and Ethics provides study resources for multiple certifications while professional organizations offer additional materials and support.

Study groups and professional networks provide peer support and additional learning opportunities through shared experiences and different perspectives. Many professionals find that group study helps them understand complex topics while maintaining motivation throughout preparation periods.

Case Example 2: The Mid-Career Certification Strategy Michael Chen worked as a hospital privacy officer for five years before pursuing CHC certification to advance his career prospects. He created a six-month study plan that included reading certification materials for one hour each morning, attending monthly study group meetings, and taking practice examinations quarterly. His systematic approach enabled him to pass the examination on his first attempt while maintaining his full-time job performance. The certification qualified him for a compliance director position at a larger health system with 25% higher compensation and broader responsibilities.

Practice examinations help candidates assess their knowledge and identify areas needing additional study. Most certification organizations provide practice tests that simulate actual examination conditions and content. Regular practice testing throughout preparation helps ensure examination readiness.

Exam Techniques and Success Factors

Certification examinations typically use multiple-choice formats that test both factual knowledge and applied understanding of compliance concepts. Successful test-taking requires both content mastery and effective examination strategies.

Time management during examinations ensures that candidates can complete all questions within allotted timeframes. Most certification examinations allow adequate time for thoughtful consideration, but candidates should practice working within time constraints.

Question analysis skills help candidates understand what examinations are asking and identify correct answers among multiple options. This includes reading questions carefully, eliminating obviously incorrect answers, and selecting best responses based on examination content.

Maintaining Certifications and Continuing Education

Certification maintenance requires ongoing education and professional development that ensures continued competency as regulations and practices evolve. Most certifications require annual continuing education credits and periodic recertification examinations or applications.

Continuing education opportunities include conferences, webinars, professional courses, and self-study programs that address current compliance topics and emerging issues. The University of Pittsburgh Law School and Compliancy Group offer various continuing education options for healthcare compliance professionals.

Professional development planning helps certified professionals maintain their credentials while advancing their careers through strategic education and experience choices. This planning should align certification maintenance with broader career development goals.

Continuing Education and Professional Organizations

Healthcare compliance requires continuous learning due to frequent regulatory changes, evolving best practices, and new compliance challenges. Professional organizations provide education, networking, and career development opportunities that support lifelong learning in healthcare compliance.

Professional Organizations and Membership Benefits

Healthcare compliance professional organizations offer specialized resources, education programs, and networking opportunities that support career development and professional excellence. Multiple organizations serve different aspects of healthcare compliance.

The Health Care Compliance Association (HCCA) represents the largest organization specifically focused on healthcare compliance, offering certifications, conferences, publications, and networking opportunities. HCCA membership provides access to regulatory updates, best practices, and peer networks that support professional development.

The Society of Corporate Compliance and Ethics (SCCE) addresses compliance across multiple industries while offering healthcare-specific resources and programs. SCCE provides certifications, education programs, and publications that address both general compliance principles and healthcare applications.

The American Health Lawyers Association (AHLA) focuses on legal aspects of healthcare including compliance, regulatory requirements, and enforcement activities. AHLA membership provides access to legal analysis, regulatory updates, and specialized education programs for healthcare professionals.

Professional organization benefits include continuing education programs, networking opportunities, career development resources, and advocacy for professional interests. Active participation in professional organizations often provides career advantages through knowledge sharing, mentorship, and job opportunities.

Conference and Training Opportunities

Professional conferences provide intensive learning opportunities through education sessions, networking events, and vendor exhibitions that showcase new products and services. Major healthcare compliance conferences typically offer multiple educational tracks addressing different specializations and experience levels.

National conferences bring together hundreds or thousands of compliance professionals for comprehensive education programs covering current topics, emerging issues, and best practices. These events provide excellent networking opportunities while offering convenient access to multiple education sessions.

Regional conferences offer more accessible alternatives to national events while providing focused education on topics relevant to specific geographic areas or organizational types. Regional events often feature local speakers and case studies that address area-specific compliance challenges.

Online Learning Resources and Flexibility

Online education programs provide flexible alternatives to in-person training while offering access to specialized topics and expert instruction. These programs are particularly valuable for professionals in rural areas or those with scheduling constraints that make conference attendance difficult.

Webinar series address current topics through short, focused sessions that provide targeted education without requiring extensive time commitments. Many professional organizations offer regular webinar programs that address emerging issues and regulatory changes.

Self-paced learning modules allow professionals to complete education programs according to their schedules while ensuring comprehensive coverage of required topics. These modules often include interactive elements, assessments, and certificates of completion.

Case Example 3: The Rural Hospital Compliance Officer's Professional Development Jennifer Adams worked as the sole compliance officer for a 75-bed rural hospital in Montana, managing all compliance functions with limited local resources. She joined HCCA and SCCE to access professional education and peer networks that weren't available locally. Through online education programs, she earned her CHC certification while maintaining her full-time responsibilities. Virtual conference participation enabled her to attend sessions on emerging compliance topics without travel expenses. Professional organization networking helped her connect with peers facing similar challenges, sharing solutions and best practices through online forums and regional meetings. This professional development supported her effectiveness in the isolated role while preparing her for future advancement opportunities.

Advanced Degrees and Education

Advanced education can accelerate healthcare compliance career development while providing deeper knowledge of healthcare operations, regulatory frameworks, and business management. Multiple degree options serve different career paths and professional interests.

Graduate Degree Options and Career Benefits

Master's degree programs provide advanced knowledge that supports healthcare compliance leadership roles and career advancement. Different degree types offer varying benefits depending on career goals and current background.

Master of Health Administration (MHA) programs focus on healthcare management, operations, and policy, providing broad understanding of healthcare delivery systems. These programs are particularly valuable for compliance professionals seeking senior leadership roles that require business knowledge alongside compliance expertise.

Master of Business Administration (MBA) programs with healthcare concentrations combine general business knowledge with healthcare-specific applications. MBA programs provide financial, strategic, and operational knowledge that supports executive-level compliance roles.

Juris Doctor (JD) programs provide legal education that supports compliance professionals working in highly regulated areas or seeking roles that require legal analysis. Law degrees are particularly valuable for fraud and abuse specialists, regulatory affairs professionals, and senior compliance executives.

Master of Science programs in health informatics, quality management, or related fields provide specialized knowledge that supports compliance roles at the intersection of technology, quality, and regulatory requirements.

Certificate Programs and Professional Development

Certificate programs offer focused education in specific compliance areas without requiring full degree commitments. These programs provide targeted knowledge development while maintaining career continuity.

Drexel Online and other institutions offer healthcare compliance certificate programs that address regulatory requirements, audit procedures, and program management. These programs typically require several months to complete while providing practical knowledge for immediate application.

University-based programs often provide continuing education units that support certification maintenance while offering academic rigor and expert instruction. Kent State University and USC Gould School of Law offer respected certificate programs in healthcare compliance.

Seton Hall Law School and other institutions provide specialized programs addressing specific compliance areas such as privacy, fraud prevention, or research compliance. These programs serve professionals seeking deep expertise in particular domains.

Executive Education and Leadership Development

Executive education programs provide intensive learning experiences for senior compliance professionals and those preparing for leadership roles. These programs typically focus on strategic thinking, leadership skills, and business acumen rather than technical compliance knowledge.

Healthcare management programs address industry-specific leadership challenges including physician relations, quality improvement, and regulatory relationships. These programs help compliance professionals develop skills needed for senior executive roles.

General management programs provide broader business knowledge including finance, strategy, human resources, and operations management. This knowledge supports compliance professionals seeking roles with broader organizational responsibilities.

Return on Investment Considerations

Advanced education represents significant investments of time and money that professionals should evaluate carefully against career goals and potential benefits. Different programs offer varying returns depending on individual circumstances and career objectives.

Salary impact studies from the University of Pittsburgh Law School show that healthcare compliance professionals with advanced degrees typically earn 15-30% more than those with bachelor's degrees, though individual results vary based on experience, location, and organizational factors.

Career advancement opportunities often require advanced education for senior roles, particularly chief compliance officer positions that typically prefer master's degrees or equivalent experience. Advanced education can accelerate advancement timelines while providing knowledge needed for leadership success.

Personal satisfaction and professional credibility benefits may justify education investments even when direct financial returns are limited. Many professionals find that advanced education enhances their effectiveness and job satisfaction regardless of immediate career benefits.

Networking and Relationship Building

Professional relationships provide career opportunities, knowledge sharing, and support systems that enhance career development throughout healthcare compliance careers. Effective networking requires systematic approaches that build authentic relationships rather than just collecting contacts.

Strategic Networking Approaches

Successful networking focuses on building mutually beneficial relationships through shared interests, professional collaboration, and knowledge exchange. The most effective networkers provide value to others rather than just seeking personal benefits.

Professional organization participation provides structured networking opportunities through committees, special interest groups, and local chapters. Active involvement in organizational activities builds deeper relationships than passive membership.

Mentorship relationships offer valuable guidance and support for career development while providing opportunities to mentor others as careers progress. Effective mentorship involves both receiving guidance from senior professionals and providing support to newer professionals.

Industry relationships with vendors, consultants, and peers at other organizations provide broader perspectives on compliance challenges and solutions. These relationships often provide job opportunities and professional references.

Online Networking and Social Media

Digital networking platforms extend relationship-building opportunities beyond geographic limitations while providing access to broader professional communities. LinkedIn and professional forums offer platforms for sharing knowledge and building relationships.

Content sharing through articles, posts, and comments demonstrates expertise while providing value to professional networks. Thoughtful contributions to professional discussions build credibility and visibility within healthcare compliance communities.

Professional online presence through LinkedIn profiles, organization websites, and professional directories helps establish credibility while making networking connections more meaningful and lasting.

Professional Growth and Excellence

Healthcare compliance career development requires continuous learning, relationship building, and strategic planning that balances personal interests with market opportunities. The most successful professionals treat their careers as long-term journeys rather than series of jobs.

The healthcare compliance field continues evolving as regulations change, technology advances, and organizational needs develop.

Professionals who embrace this evolution while maintaining strong foundations in compliance fundamentals achieve the greatest career success and professional satisfaction.

Career satisfaction in healthcare compliance often comes from meaningful work that protects patients while supporting organizational mission. The combination of regulatory expertise, healthcare knowledge, and business acumen creates opportunities to make real differences in healthcare quality and safety.

Professional Excellence Principles

- Career development in healthcare compliance requires systematic planning that balances specialization with broad business knowledge and leadership skills
- Professional certifications provide competitive advantages and validate expertise, but require ongoing maintenance through continuing education and professional development
- Professional organizations offer essential resources for networking, education, and career advancement that support lifelong learning in a rapidly changing field
- Advanced education can accelerate career development and provide knowledge needed for senior roles, but should be evaluated carefully against career goals and financial considerations
- Effective networking builds authentic relationships that provide career opportunities, knowledge sharing, and professional support throughout healthcare compliance careers

Chapter 12: Practical Implementation Guide

Starting a new role as a healthcare compliance officer can feel overwhelming given the breadth of responsibilities, complexity of regulations, and organizational expectations for immediate impact. Success requires systematic approaches that balance urgent priorities with long-term program development while building relationships and credibility throughout the organization. This practical guide provides actionable strategies for compliance officers at any stage of their careers.

Day One as a Compliance Officer

Your first days as a healthcare compliance officer set the tone for your entire tenure while providing opportunities to assess organizational needs, establish priorities, and begin building relationships that will determine your long-term success. These initial activities require careful balance between listening and leading, assessment and action.

Initial Assessment Strategies and Information Gathering

Your initial assessment should provide comprehensive understanding of organizational compliance status, risk profile, and cultural factors that affect program effectiveness. This assessment guides priority setting while identifying quick wins that demonstrate value immediately.

Document review begins with existing compliance policies, audit reports, investigation files, and regulatory correspondence that reveal current program status and historical compliance performance. Look for patterns in findings, recurring problems, and areas with recent regulatory changes. This review provides baseline understanding while identifying immediate concerns.

Stakeholder interviews with senior leadership, department heads, and front-line staff reveal organizational culture, compliance challenges, and improvement opportunities. These conversations help you understand how compliance is perceived throughout the organization while building relationships with key partners.

Regulatory assessment examines current requirements affecting your organization, recent enforcement actions in your industry, and emerging regulatory trends that might create future compliance challenges. This assessment helps you understand the regulatory environment while identifying areas needing attention.

Case Example 1: The New CCO's Strategic Assessment When David Park became Chief Compliance Officer at a 300-bed community hospital, he spent his first month conducting structured interviews with all department heads, reviewing three years of audit reports, and analyzing regulatory enforcement trends affecting similar organizations. His assessment revealed that while the organization had strong clinical quality programs, billing compliance oversight was inadequate and privacy incident response was inconsistent. This systematic assessment enabled him to present a focused improvement plan to the board within 45 days, demonstrating both thoroughness and urgency.

System evaluation includes information technology capabilities, reporting mechanisms, and data sources that support compliance monitoring and reporting. Understanding these systems helps you identify capability gaps while leveraging existing resources effectively.

Benchmarking research compares your organization's compliance performance with peer organizations, industry standards, and best practices. This comparison helps identify performance gaps while setting realistic improvement targets.

Quick Wins Identification and Early Impact

Quick wins demonstrate your value while building momentum for larger improvement initiatives. These early successes should address real compliance risks while being achievable within your first few months.

Low-hanging fruit includes compliance problems that can be resolved quickly with minimal resources or organizational change. This might include updating outdated policies, completing overdue training, or fixing obvious system configuration problems.

Visible improvements in areas that affect daily operations help staff understand compliance benefits while building support for your program. Streamlining approval processes, clarifying confusing procedures, or resolving persistent problems demonstrates practical value.

Communication enhancements including regular updates, clear reporting, and accessible guidance help staff understand compliance requirements while demonstrating your commitment to supporting rather than policing their work.

Stakeholder Engagement and Relationship Building

Successful compliance officers build collaborative relationships rather than adversarial ones, positioning themselves as partners who help achieve organizational goals rather than obstacles that create additional work. These relationships require ongoing attention and genuine commitment to supporting others' success.

Leadership alignment ensures that senior executives understand your approach, support your priorities, and provide necessary resources for program success. Regular communication with CEO, board members, and other executives builds confidence while ensuring strategic alignment.

Department partnerships with clinical, financial, and operational leaders create collaborative approaches to compliance challenges.

These partnerships help you understand department-specific needs while ensuring compliance solutions support rather than interfere with operational effectiveness.

Physician engagement requires special attention because physician buy-in significantly affects compliance program success. Physicians respond well to evidence-based approaches, clinical examples, and solutions that improve rather than complicate patient care.

Priority Setting and Resource Allocation

Effective priority setting balances urgent compliance needs with strategic program development while considering available resources and organizational capacity for change. Clear priorities help you focus efforts while communicating expectations to stakeholders.

Risk-based prioritization focuses on areas with highest probability and impact of compliance violations while considering organizational capacity to address multiple issues simultaneously. This approach ensures you address the most serious risks first while building program credibility.

Regulatory deadlines and mandatory requirements must be addressed regardless of other priorities, but can often be integrated with broader improvement initiatives that create additional value. Look for opportunities to exceed minimum requirements while building program capabilities.

Case Example 2: The Physician Practice Compliance Officer's Priorities Maria Rodriguez joined a large cardiology practice as their first dedicated compliance officer after previous compliance violations resulted in a Corporate Integrity Agreement. Her initial assessment revealed multiple compliance gaps, but she prioritized billing compliance improvements that were required by the CIA while integrating broader program development. She implemented systematic chart audits, enhanced physician documentation training, and created regular compliance reporting—all CIA requirements that also strengthened the overall compliance program. This integrated approach satisfied regulatory requirements while building sustainable program capabilities.

Common Challenges and Solutions

Healthcare compliance officers face predictable challenges that require systematic solutions rather than ad hoc responses. Understanding these challenges and proven solutions helps you prepare for common problems while developing resilience for unexpected situations.

Resource Constraints and Budget Limitations

Limited resources represent the most common challenge facing healthcare compliance officers, requiring creative approaches that maximize impact while working within budget constraints. Successful compliance officers learn to leverage existing resources while building business cases for additional investments.

Staff augmentation through cross-training existing employees, engaging temporary resources for specific projects, or partnering with other departments can expand your effective capacity without increasing permanent headcount. Look for opportunities to develop compliance expertise throughout the organization rather than concentrating all functions in your department.

Technology leverage includes maximizing existing system capabilities, negotiating favorable vendor arrangements, and implementing low-cost solutions that provide high value. Many organizations underutilize technology capabilities they already own while purchasing expensive solutions for problems that could be solved with existing resources.

Phased implementation allows you to address compliance needs systematically without overwhelming organizational capacity or budget. Breaking large projects into manageable phases demonstrates progress while allowing time to secure additional resources as programs prove their value.

Partnership development with other healthcare organizations, shared service arrangements, or consulting relationships can provide access to expertise and resources that individual organizations cannot afford independently. Regional collaboratives and system affiliations often provide cost-effective approaches to specialized compliance needs.

Cultural Resistance and Change Management

Organizational culture significantly affects compliance program success, with resistant cultures undermining even well-designed programs while supportive cultures enabling success despite resource limitations. Understanding and addressing cultural factors requires patience, persistence, and strategic approaches.

Communication strategies should emphasize how compliance supports rather than interferes with organizational mission and individual job satisfaction. Frame compliance as patient protection rather than regulatory burden, demonstrating connections between requirements and quality care.

Education programs that address underlying concerns and misconceptions help staff understand why compliance matters while providing practical guidance for meeting requirements. Focus on real-world examples and practical applications rather than abstract regulatory concepts.

Incentive alignment includes recognizing and rewarding compliance excellence while ensuring that performance evaluation and compensation systems support rather than undermine compliance objectives. Mixed messages from leadership undermine compliance programs regardless of policy statements.

Case Example 3: The Health System Cultural Transformation
Regional Health System's new Chief Compliance Officer found

widespread resistance to compliance initiatives, with staff viewing compliance as bureaucratic interference rather than patient protection. She addressed this resistance through story-telling that connected compliance requirements to patient safety outcomes, recognizing departments that demonstrated compliance excellence, and ensuring that operational leaders reinforced compliance messages. Over 18 months, employee compliance perception scores improved from 2.3 to 4.1 on a 5-point scale, and compliance performance improved across all measured areas.

Leadership support must be visible and consistent, with executives demonstrating compliance commitment through their words and actions. Mixed messages from leadership quickly undermine compliance programs regardless of formal policy statements.

Technology Limitations and System Integration

Healthcare organizations often operate with outdated or fragmented technology systems that make compliance monitoring and reporting difficult. Working within these limitations while advocating for improvements requires both technical understanding and business acumen.

Workaround development helps you achieve compliance objectives despite system limitations through manual processes, data exports, or alternative approaches that provide necessary information and controls. While not ideal, effective workarounds can provide temporary solutions while building business cases for system improvements.

Vendor partnerships with existing technology suppliers can sometimes provide cost-effective enhancements or configurations that improve compliance capabilities without requiring major system replacements. Many vendors offer compliance-specific modules or services that integrate with existing platforms.

Business case development for technology improvements should quantify compliance benefits alongside operational efficiencies, risk reduction, and revenue opportunities. Technology decisions often require justification beyond compliance needs, making comprehensive benefit analysis essential.

Integration planning ensures that compliance technology improvements align with broader organizational technology strategies while considering future needs and expansion possibilities. Short-term fixes that create long-term problems should be avoided when possible.

Regulatory Complexity and Interpretation

Healthcare regulations are often complex, ambiguous, or conflicting, requiring judgment calls that can significantly affect organizational risk exposure. Managing regulatory complexity requires systematic approaches to research, analysis, and decision-making.

Legal counsel relationships provide access to specialized expertise for complex regulatory questions while ensuring appropriate legal privilege protections for sensitive matters. Establish clear guidelines for when legal consultation is needed and how decisions will be documented.

Professional network utilization through industry associations, peer organizations, and consulting relationships provides access to diverse perspectives on regulatory interpretation and implementation. Other compliance officers facing similar challenges can provide practical insights that complement legal analysis.

Documentation requirements for regulatory decisions should capture analysis conducted, alternatives considered, and rationale for decisions made. This documentation supports decision quality while providing protection if interpretations are later questioned.

Regulatory monitoring systems help you stay current with changing requirements while identifying emerging issues that might affect your organization. Systematic monitoring prevents surprises while enabling proactive planning for regulatory changes.

Best Practices from the Field

Healthcare compliance success depends on learning from others' experiences, both successes and failures. These field-tested practices provide proven approaches that can be adapted to different organizational contexts and compliance challenges.

Successful Program Examples and Lessons Learned

High-performing compliance programs share common characteristics including leadership support, cultural integration, systematic processes, and continuous improvement. Studying these programs provides insights that can be applied across different organizational settings.

Integrated delivery system programs often demonstrate how compliance can be standardized across multiple facilities while allowing local adaptation for specific needs. These programs show how consistent policies and procedures can be implemented while maintaining operational flexibility.

Academic medical center programs illustrate how compliance programs can address complex regulatory requirements including research compliance, teaching obligations, and clinical care standards simultaneously. These programs demonstrate sophisticated approaches to multi-domain compliance.

Community hospital programs show how effective compliance can be achieved with limited resources through strategic partnerships, technology leverage, and staff cross-training. These programs prove that size doesn't determine compliance effectiveness.

Enforcement Action Analysis and Prevention

Studying recent enforcement actions provides insights into regulatory priorities, common violations, and effective prevention strategies. Akerman LLP and U.S. Department of Justice enforcement data reveal patterns that inform compliance program design.

Common violation patterns include billing compliance failures, privacy breaches, quality reporting errors, and physician relationship problems. Understanding these patterns helps you focus prevention efforts on highest-risk areas.

Enforcement trends reveal regulatory priorities and emerging focus areas that might affect your organization. Recent emphasis on cybersecurity, opioid prescribing, and telehealth compliance reflects current regulatory concerns.

Prevention strategies based on enforcement analysis include enhanced monitoring in high-risk areas, proactive policy updates, and staff education about common violation patterns. Learning from others' mistakes helps you avoid similar problems.

Resolution approaches used in recent settlements provide guidance for self-disclosure decisions and corrective action planning. Understanding how similar organizations resolved compliance violations helps you prepare for potential problems.

Innovation Case Studies and Emerging Practices

Leading healthcare organizations continue innovating in compliance program design, technology utilization, and performance measurement. These innovations provide examples of advanced practices that can be adapted to other settings.

Technology innovations include artificial intelligence applications, predictive analytics, and automated monitoring that improve compliance effectiveness while reducing administrative burden. Early adopters demonstrate both benefits and challenges of new technologies.

Process innovations include risk-based auditing, integrated reporting, and cross-functional compliance teams that improve program efficiency while enhancing organizational integration. These innovations show how traditional compliance approaches can be updated for current needs.

Performance measurement innovations include balanced scorecards, predictive indicators, and stakeholder feedback systems that provide more comprehensive pictures of compliance program effectiveness. These approaches move beyond traditional compliance metrics to assess broader program impact.

Building Your Compliance Toolkit

Effective healthcare compliance requires systematic approaches supported by appropriate tools, templates, and resources. Building your toolkit systematically ensures you have necessary resources while avoiding duplicative efforts.

Essential Documents and Templates

Standard documents and templates provide consistent approaches to common compliance activities while saving time and ensuring thoroughness. These resources should be customized for your organization while maintaining essential elements.

Policy templates for common compliance areas including privacy, billing, quality reporting, and physician relationships provide starting points for policy development while ensuring coverage of essential requirements. Templates should be adapted to organizational culture and specific needs.

Audit checklists and procedures ensure systematic approaches to compliance assessment while providing documentation for regulatory requirements. Standard checklists help ensure consistency across different auditors and time periods.

Investigation protocols guide response to potential compliance violations while ensuring appropriate documentation and follow-up. Clear protocols help ensure consistent, thorough responses regardless of who conducts investigations.

Training materials including presentations, handouts, and assessments provide standardized education while allowing customization for different audiences. Effective training materials address both regulatory requirements and practical applications.

Technology Recommendations and Selection Criteria

Technology tools can significantly improve compliance program effectiveness when selected and implemented appropriately. Understanding available options helps you make informed decisions that provide good value for your organization.

Audit management systems help organize, conduct, and track compliance audits while providing consistent documentation and reporting. These systems are particularly valuable for organizations conducting multiple audits across different compliance areas.

Policy management platforms provide version control, distribution tracking, and automated review reminders that ensure policies remain current while reducing administrative burden. These platforms are essential for organizations with large numbers of policies.

Training management systems track completion, test results, and certification requirements while providing automated reminders and reporting capabilities. These systems help ensure comprehensive training coverage while reducing manual tracking requirements.

Professional Resources and Ongoing Learning

Healthcare compliance requires continuous learning due to frequent regulatory changes and evolving best practices. Building relationships with professional resources ensures you have access to current information and expert guidance.

Professional organization memberships provide access to education programs, networking opportunities, and regulatory updates that support ongoing professional development. Active participation maximizes benefits while building professional relationships.

Legal counsel relationships provide access to specialized expertise while ensuring appropriate privilege protections. Establish relationships before you need urgent legal advice to ensure effective partnerships.

Consulting relationships with specialized compliance firms can provide expertise for specific projects or ongoing support for complex compliance areas. These relationships can be particularly valuable for smaller organizations that cannot afford full-time specialized staff.

Educational resources including books, journals, and online courses provide ongoing learning opportunities while supporting certification maintenance requirements. Systematic professional development helps you stay current while advancing your career.

The Road to Excellence

Healthcare compliance implementation requires balance between immediate needs and long-term sustainability, technical requirements and organizational culture, regulatory obligations and operational effectiveness. Success comes from systematic approaches that address these tensions thoughtfully while maintaining focus on the fundamental goal of protecting patients through excellent care.

The most effective compliance officers combine technical expertise with practical wisdom, regulatory knowledge with business understanding, and personal integrity with collaborative leadership.

These combinations enable them to navigate complex challenges while building programs that genuinely protect patients and support organizational mission.

Your compliance journey will include both successes and setbacks, moments of clarity and periods of confusion, strong support and occasional resistance. The key is maintaining perspective about the importance of your work while continuously learning and improving your effectiveness. Every policy implemented, every audit conducted, and every training session delivered contributes to better patient care and organizational excellence.

Implementation Wisdom

- Successful compliance officers balance immediate needs with long-term program development through systematic assessment, priority setting, and relationship building
- Common challenges including resource constraints, cultural resistance, and technology limitations require systematic solutions rather than ad hoc responses
- Best practices from high-performing organizations provide proven approaches that can be adapted to different organizational contexts and compliance challenges
- Building comprehensive compliance toolkits with essential documents, appropriate technology, and professional resources supports ongoing program effectiveness
- Long-term success requires continuous learning, relationship building, and commitment to the fundamental goal of protecting patients through excellent healthcare delivery

Chapter 13: Practice Questions

Healthcare Compliance Program Fundamentals

1. According to OIG guidance, how many elements comprise an effective healthcare compliance program? A) Five elements B) Six elements C) Seven elements D) Eight elements

Answer: C) Seven elements. The Office of Inspector General established seven fundamental elements for effective compliance programs: written policies and procedures, compliance officer and committee oversight, effective training and education, communication systems, monitoring and auditing, response to violations, and enforcement of disciplinary standards.

2. Who should the chief compliance officer report to in order to maintain independence? A) Chief Financial Officer B) Chief Operating Officer C) Chief Executive Officer or Board of Directors D) Medical Staff President

Answer: C) Chief Executive Officer or Board of Directors. Independence is essential for compliance effectiveness. Direct reporting to the CEO or board ensures the compliance officer can address issues without fear of retaliation and has access to necessary resources and authority.

3. What is the minimum frequency for compliance committee meetings in most healthcare organizations? A) Weekly B) Monthly C) Quarterly D) Annually

Answer: C) Quarterly. Most healthcare organizations conduct compliance committee meetings quarterly to ensure regular oversight while allowing sufficient time between meetings to implement and assess corrective actions.

4. Which document establishes the ethical framework for organizational behavior? A) Compliance policies B) Standards of conduct C) Training materials D) Audit procedures

Answer: B) Standards of conduct. Standards of conduct communicate organizational values and expectations for ethical behavior, serving as the foundational document that guides decision-making throughout the organization.

5. What type of training is most effective for healthcare compliance programs? A) Annual general sessions only B) Online modules only C) Role-specific training tailored to job functions D) Mandatory lectures for all staff

Answer: C) Role-specific training tailored to job functions. Effective compliance training addresses specific risks and requirements relevant to each employee's responsibilities, making the content more meaningful and actionable than generic training approaches.

6. How often should compliance policies be reviewed and updated? A) Every five years B) Every three years C) Annually or when regulations change D) Only when violations occur

Answer: C) Annually or when regulations change. Healthcare regulations change frequently, requiring regular policy review to ensure continued accuracy and effectiveness. Annual reviews help maintain current policies while regulatory changes may necessitate immediate updates.

7. What is the primary purpose of a compliance hotline? A) To report financial problems B) To provide anonymous reporting of potential compliance violations C) To handle patient complaints D) To manage vendor relationships

Answer: B) To provide anonymous reporting of potential compliance violations. Compliance hotlines enable employees to report suspected violations anonymously, encouraging reporting by removing fear of retaliation while providing early warning of potential compliance problems.

8. Which factor is most important when selecting compliance committee members? A) Seniority within the organization B) Representation from major operational areas C) Previous compliance experience D) Educational background

Answer: B) Representation from major operational areas. Effective compliance committees include representatives from all major organizational functions to ensure comprehensive oversight and understanding of compliance risks across different areas.

9. What should be the primary focus when developing corrective action plans? A) Punishing responsible individuals B) Addressing root causes of problems C) Meeting regulatory deadlines D) Minimizing costs

Answer: B) Addressing root causes of problems. Effective corrective action plans address underlying system problems that allowed violations to occur, preventing recurrence rather than just responding to symptoms of deeper issues.

10. How should compliance effectiveness be measured? A) Number of policies created B) Training completion rates only C) Multiple metrics including process and outcome measures D) Absence of regulatory violations

Answer: C) Multiple metrics including process and outcome measures. Comprehensive compliance measurement includes both leading indicators like training completion and audit results, and lagging indicators like violation rates and regulatory feedback.

11. What is the most appropriate response when a compliance violation is discovered? A) Immediate termination of responsible parties B) Cover up to protect organizational reputation C) Systematic investigation and appropriate corrective action D) Wait to see if regulators discover the problem

Answer: C) Systematic investigation and appropriate corrective action. Discovered violations require thorough investigation to understand scope and causes, followed by appropriate corrective actions that address both the immediate problem and underlying issues.

12. Which approach is most effective for building a culture of compliance? A) Strict enforcement and punishment B) Leadership modeling and positive reinforcement C) Detailed policies and procedures D) Frequent auditing and monitoring

Answer: B) Leadership modeling and positive reinforcement. Organizational culture is primarily influenced by leadership behavior and reinforcement systems. Leaders who consistently demonstrate compliance commitment create cultures that value ethical behavior.

13. What is the primary benefit of having written compliance policies? A) Meeting regulatory requirements B) Providing legal protection C) Guiding consistent decision-making D) Documenting organizational structure

Answer: C) Guiding consistent decision-making. While policies serve multiple purposes, their primary benefit is providing clear guidance that enables consistent, appropriate decision-making throughout the organization.

14. How should compliance officers handle conflicts between compliance requirements and operational pressures? A) Always prioritize operational needs B) Always prioritize compliance requirements C) Seek solutions that address both concerns D) Escalate all conflicts to senior leadership

Answer: C) Seek solutions that address both concerns. Effective compliance officers work collaboratively to find solutions that meet compliance requirements while supporting operational effectiveness, rather than creating adversarial relationships.

15. What is the most important factor in compliance training effectiveness? A) Length of training sessions B) Frequency of training delivery C) Relevance to job responsibilities D) Use of multimedia presentations

Answer: C) Relevance to job responsibilities. Training effectiveness depends primarily on relevance to participants' daily work. Training that addresses specific job-related compliance risks is more likely to change behavior than general compliance education.

16. Which communication method is most effective for compliance program promotion? A) Annual compliance newsletters B) Multiple channels with consistent messaging C) Email announcements only D) Posting notices in common areas

Answer: B) Multiple channels with consistent messaging. Effective communication uses various channels to reach different audiences while maintaining consistent messages. Multiple touchpoints increase the likelihood that important information reaches all staff members.

17. What is the primary purpose of compliance monitoring and auditing? A) Finding violations to report B) Demonstrating regulatory compliance C) Early detection and prevention of problems D) Satisfying external audit requirements

Answer: C) Early detection and prevention of problems. While monitoring serves multiple purposes, its primary value is identifying potential problems before they become serious violations, enabling proactive correction rather than reactive response.

18. How should organizations respond to employee concerns about potential retaliation for compliance reporting? A) Guarantee no consequences will occur B) Implement strong non-retaliation policies with enforcement C) Encourage anonymous reporting only D) Limit reporting to senior management

Answer: B) Implement strong non-retaliation policies with enforcement. Effective non-retaliation protection requires clear policies, consistent enforcement, and demonstrated organizational commitment to protecting employees who report concerns in good faith.

19. What is the most critical element of an effective compliance program? A) Detailed written policies B) Sophisticated monitoring systems C) Leadership commitment and support D) Comprehensive training programs

Answer: C) Leadership commitment and support. While all elements are important, leadership commitment provides the foundation that enables all other program elements to function effectively. Without leadership support, other elements cannot succeed.

20. Which factor most influences employee compliance behavior? A) Fear of punishment B) Understanding of requirements C) Organizational culture and expectations D) Regulatory enforcement actions

Answer: C) Organizational culture and expectations. Employee behavior is primarily influenced by organizational culture and peer expectations. Understanding requirements is necessary but not sufficient without cultural support for compliance.

21. What is the best approach for handling compliance questions from staff? A) Refer all questions to legal counsel B) Provide immediate answers based on personal knowledge C) Research thoroughly and provide accurate guidance D) Direct staff to find answers independently

Answer: C) Research thoroughly and provide accurate guidance. Compliance officers should provide accurate, well-researched guidance rather than quick answers that might be incorrect. Taking time to research ensures quality advice while building credibility.

22. How should compliance officers prioritize their activities? A) Focus only on regulatory requirements B) Address the loudest complaints first C) Use risk-based approaches considering likelihood and impact D) Rotate equally among all compliance areas

Answer: C) Use risk-based approaches considering likelihood and impact. Risk-based prioritization ensures that limited resources focus on areas with the greatest potential for harm, considering both the probability of problems and their potential consequences.

23. What is the most effective way to demonstrate compliance program value? A) Count the number of policies created B) Track training completion percentages C) Measure problem prevention and risk reduction D) Report regulatory inspection results

Answer: C) Measure problem prevention and risk reduction. Program value is best demonstrated through prevented problems, reduced risks, and improved outcomes rather than activity measures like policies created or training completed.

24. Which approach is most effective for policy development? A) Copy policies from other organizations B) Involve relevant stakeholders in development C) Create comprehensive policies covering all possibilities D) Keep policies brief and general

Answer: B) Involve relevant stakeholders in development. Stakeholder involvement ensures policies are practical, understandable, and implementable while building buy-in from those who must follow the policies.

25. What is the primary purpose of compliance documentation? A) Meeting regulatory requirements B) Providing legal protection C)

Supporting accountability and improvement D) Demonstrating activity levels

Answer: C) Supporting accountability and improvement. While documentation serves multiple purposes, its primary value is creating accountability for compliance activities and providing information needed for continuous improvement.

26. How should compliance officers handle situations where regulations are unclear? A) Wait for regulatory clarification B) Choose the most conservative interpretation C) Seek expert guidance and document decisions D) Ignore unclear requirements

Answer: C) Seek expert guidance and document decisions. Unclear regulations require careful analysis, expert consultation, and documented decision-making processes that can withstand later scrutiny if interpretations are questioned.

27. What is the most important characteristic of effective compliance leadership? A) Technical expertise in all regulations B) Ability to build relationships and influence others C) Experience in healthcare operations D) Legal background and training

Answer: B) Ability to build relationships and influence others. Compliance officers must influence behavior across organizations without direct authority over most staff. Relationship-building and influence skills are essential for success.

28. Which metric best indicates compliance program effectiveness? A) Number of violations found B) Training completion rates C) Trend improvements in key risk areas D) Audit scores from external reviewers

Answer: C) Trend improvements in key risk areas. Effectiveness is best measured by sustained improvements in compliance performance over time, indicating that the program is successfully reducing risks and preventing problems.

29. What is the most appropriate role for compliance officers in organizational decision-making? A) Making all decisions with compliance implications B) Providing compliance input and guidance C) Approving decisions after they are made D) Avoiding involvement in business decisions

Answer: B) Providing compliance input and guidance. Compliance officers should provide expert guidance and input during decision-making processes while allowing business leaders to make informed decisions that consider compliance alongside other factors.

30. How should compliance programs address resource constraints? A) Reduce program scope to match resources B) Request unlimited additional resources C) Prioritize activities based on risk and impact D) Maintain all activities regardless of quality

Answer: C) Prioritize activities based on risk and impact. Resource constraints require strategic prioritization that focuses available resources on highest-risk areas while maintaining adequate coverage of all compliance domains.

31. What is the best approach for measuring compliance training effectiveness? A) Test scores immediately after training B) Training completion percentages C) Behavior change and performance improvement D) Participant satisfaction surveys

Answer: C) Behavior change and performance improvement. Training effectiveness should be measured by actual behavior change and performance improvement rather than completion rates or immediate test scores that may not predict lasting change.

32. Which factor is most important when designing compliance monitoring systems? A) Comprehensive coverage of all activities B) Focus on highest-risk areas and activities C) Equal attention to all compliance domains D) Maximum automation and efficiency

Answer: B) Focus on highest-risk areas and activities. Effective monitoring systems focus on areas with greatest risk exposure while maintaining adequate coverage of other areas. Risk-based approaches maximize the value of limited monitoring resources.

33. How should compliance officers handle pressure to overlook potential violations? A) Always escalate to senior leadership B) Comply with organizational priorities C) Maintain independence and investigate appropriately D) Document concerns but take no action

Answer: C) Maintain independence and investigate appropriately. Compliance officer independence requires appropriate investigation of potential violations regardless of organizational pressure. Independence is essential for program credibility and effectiveness.

34. What is the most effective approach for compliance policy implementation? A) Distribute policies and require acknowledgment B) Provide training and support during implementation C) Monitor compliance through frequent audits D) Enforce policies through disciplinary action

Answer: B) Provide training and support during implementation. Successful policy implementation requires training, support, and assistance during the transition period rather than just distribution and enforcement.

35. Which communication approach is most effective for compliance messages? A) Focus on regulatory requirements and penalties B) Emphasize patient safety and quality care C) Highlight organizational risks and liability D) Stress individual responsibility and accountability

Answer: B) Emphasize patient safety and quality care. Messages that connect compliance requirements to patient safety and quality care resonate better with healthcare workers than those focusing solely on regulatory or legal concerns.

36. What is the primary benefit of compliance benchmarking? A) Demonstrating superior performance B) Identifying improvement opportunities C) Meeting regulatory requirements D) Reducing program costs

Answer: B) Identifying improvement opportunities. Benchmarking helps organizations understand their performance relative to peers and identify areas where improvement is possible, providing roadmaps for program enhancement.

37. How should compliance officers balance thoroughness with efficiency? A) Always prioritize thoroughness over efficiency B) Always prioritize efficiency over thoroughness C) Use risk-based approaches to optimize both D) Let senior leadership make these decisions

Answer: C) Use risk-based approaches to optimize both. Risk-based approaches enable appropriate thoroughness in high-risk areas while maintaining efficiency in lower-risk activities, optimizing the use of available resources.

38. What is the most important factor in compliance investigation success? A) Speed of investigation completion B) Objectivity and thoroughness of analysis C) Minimizing organizational disruption D) Reaching predetermined conclusions

Answer: B) Objectivity and thoroughness of analysis. Effective investigations require objective analysis and thorough fact-gathering to reach accurate conclusions that can withstand scrutiny and support appropriate corrective actions.

39. Which approach is most effective for building compliance accountability? A) Punitive measures for all violations B) Performance measurement and feedback systems C) Detailed policies and procedures D) Frequent compliance training sessions

Answer: B) Performance measurement and feedback systems. Accountability develops through performance measurement systems that provide regular feedback and recognition, creating positive reinforcement for compliance excellence.

40. What is the best strategy for managing compliance in multi-site organizations? A) Centralized control of all compliance activities B) Complete decentralization to local sites C) Standardized policies with local implementation support D) Independent compliance programs at each site

Answer: C) Standardized policies with local implementation support. Multi-site organizations benefit from consistent policies and standards while providing local support for implementation that addresses site-specific needs and challenges.

41. How should compliance officers approach continuous improvement? A) Focus on correcting identified problems only B) Systematically evaluate and enhance all program elements C) Wait for regulatory changes to drive improvements D) Benchmark against minimum regulatory requirements

Answer: B) Systematically evaluate and enhance all program elements. Continuous improvement requires systematic evaluation of all program elements, seeking opportunities for enhancement beyond just correcting identified problems.

42. What is the most effective way to engage physicians in compliance activities? A) Mandate participation through medical staff bylaws B) Focus on clinical quality and patient safety connections C) Emphasize legal and regulatory requirements D) Provide financial incentives for participation

Answer: B) Focus on clinical quality and patient safety connections. Physicians respond best to compliance messages that connect requirements to clinical quality and patient safety rather than regulatory or legal concerns.

43. Which factor most influences compliance program sustainability? A) Adequate funding and resources B) Leadership continuity and support C) Regulatory stability and clarity D) Staff expertise and training

Answer: B) Leadership continuity and support. Sustainable compliance programs require consistent leadership support over time. Changes in leadership can undermine even well-established programs without continued commitment.

44. What is the best approach for handling compliance program challenges? A) Escalate all challenges to senior leadership B) Address challenges independently without assistance C) Seek collaborative solutions with stakeholders D) Document challenges but continue current approaches

Answer: C) Seek collaborative solutions with stakeholders. Collaborative problem-solving engages stakeholders in finding solutions while building support for compliance activities and improving program effectiveness.

45. How should compliance officers measure their personal effectiveness? A) Number of violations prevented B) Organizational compliance performance improvements C) Recognition from senior leadership D) Completion of planned activities

Answer: B) Organizational compliance performance improvements. Personal effectiveness is best measured by improvements in organizational compliance performance rather than activity levels or recognition received.

46. What is the primary goal of compliance risk assessment? A) Documenting potential liability exposure B) Identifying and prioritizing improvement opportunities C) Meeting regulatory requirements D) Preparing for external audits

Answer: B) Identifying and prioritizing improvement opportunities. Risk assessment should identify and prioritize areas where compliance improvements can reduce organizational exposure and improve performance.

47. Which approach is most effective for compliance policy communication? A) Email distribution with read receipts B) Multiple channels with reinforcement messaging C) Annual policy review sessions D) Posting on organizational intranets

Answer: B) Multiple channels with reinforcement messaging. Effective policy communication uses multiple channels and repeated reinforcement to ensure messages reach all intended recipients and are understood correctly.

48. What is the best way to evaluate compliance training needs? A) Survey employees about their interests B) Analyze compliance performance and risk data C) Follow regulatory training requirements D) Benchmark against other organizations

Answer: B) Analyze compliance performance and risk data. Training needs assessment should be based on objective analysis of compliance performance and risk exposure rather than subjective preferences or external benchmarks.

49. How should compliance officers handle competing organizational priorities? A) Always prioritize compliance over other concerns B) Defer to operational leadership decisions C) Seek solutions that address multiple priorities D) Escalate all conflicts to senior leadership

Answer: C) Seek solutions that address multiple priorities. Effective compliance officers work collaboratively to find solutions that meet compliance requirements while supporting other organizational priorities rather than creating conflicts.

50. What is the most important outcome of an effective compliance program? A) Absence of regulatory violations B) High scores on compliance assessments C) Protection of patients and organizational mission D) Positive relationships with regulators

Answer: C) Protection of patients and organizational mission. While avoiding violations and maintaining good regulatory relationships are important, the ultimate goal is protecting patients and supporting the organizational mission through ethical, compliant operations.

HIPAA Privacy and Security

51. What does PHI stand for in HIPAA terminology? A) Personal Health Information B) Protected Health Information C) Private Health Information D) Public Health Information

Answer: B) Protected Health Information. Protected Health Information (PHI) is the correct terminology under HIPAA, referring to individually identifiable health information held or transmitted by covered entities and their business associates.

52. How many identifiers must be removed under the Safe Harbor method of de-identification? A) 16 identifiers B) 17 identifiers C) 18 identifiers D) 19 identifiers

Answer: C) 18 identifiers. The Safe Harbor method requires removal of 18 specific identifiers including names, addresses, telephone numbers, email addresses, social security numbers, medical record numbers, and others.

53. What is the maximum penalty for a single HIPAA violation under current penalty structure? A) $1.5 million B) $1.8 million C) $2.0 million D) $2.2 million

Answer: B) $1.8 million. Current HIPAA penalty tiers can reach $1.8 million for a single violation in cases involving willful neglect that is not corrected within 30 days.

54. How long do covered entities have to respond to individual access requests? A) 15 days B) 30 days C) 45 days D) 60 days

Answer: B) 30 days. Covered entities must respond to individual access requests within 30 days, with one possible 30-day extension if additional time is needed and the individual is notified.

55. What is required for a breach under HIPAA? A) Intentional disclosure of PHI B) Unauthorized acquisition, access, use, or disclosure of PHI C) Criminal misuse of health information D) Financial harm to individuals

Answer: B) Unauthorized acquisition, access, use, or disclosure of PHI. A breach is defined as unauthorized acquisition, access, use, or disclosure of PHI that compromises the security or privacy of the information, regardless of intent or harm.

56. When must covered entities notify OCR of breaches affecting 500 or more individuals? A) Immediately upon discovery B) Within 30 days of discovery C) Within 60 days of discovery D) Within 90 days of discovery

Answer: C) Within 60 days of discovery. Breaches affecting 500 or more individuals must be reported to OCR within 60 days of discovery, while smaller breaches are reported annually.

57. What is the minimum necessary standard under HIPAA? A) Use the least amount of PHI possible B) Limit PHI to the minimum reasonably necessary for the purpose C) Obtain individual authorization for all uses D) Encrypt all PHI transmissions

Answer: B) Limit PHI to the minimum reasonably necessary for the purpose. The minimum necessary standard requires limiting PHI uses and disclosures to the minimum amount reasonably necessary to accomplish the intended purpose.

58. Which of the following is NOT a required HIPAA Security Rule safeguard category? A) Administrative safeguards B) Physical safeguards C) Technical safeguards D) Procedural safeguards

Answer: D) Procedural safeguards. HIPAA Security Rule requires three categories of safeguards: administrative, physical, and technical. Procedural safeguards are not a separate category but are incorporated within the three required categories.

59. What is required in a business associate agreement? A) Specification of permitted uses and disclosures B) Requirement for appropriate safeguards C) Procedures for reporting security incidents D) All of the above

Answer: D) All of the above. Business associate agreements must specify permitted uses and disclosures, require appropriate safeguards, include procedures for reporting security incidents, and address several other required elements.

60. How often must covered entities conduct risk assessments under the Security Rule? A) Annually B) Every two years C) When required by changes or incidents D) No specific timeframe is required

Answer: D) No specific timeframe is required. HIPAA Security Rule requires risk assessments but does not specify frequency. However, assessments should be conducted regularly and when significant changes occur to systems or threats.

61. What constitutes a "disclosure" under HIPAA? A) Use of PHI within the covered entity B) Release of PHI outside the covered entity C) Access to PHI by authorized personnel D) Storage of PHI in electronic systems

Answer: B) Release of PHI outside the covered entity. Disclosure specifically refers to the release, transfer, or provision of access to PHI outside the covered entity, while use refers to PHI handling within the entity.

62. Which individuals have the right to access their PHI? A) Only adult patients B) Adult patients and parents of minor children C) Individuals or their personal representatives D) Anyone with written authorization

Answer: C) Individuals or their personal representatives. The right of access belongs to individuals or their personal representatives, which may include parents of minor children, guardians, or others legally authorized to act on behalf of the individual.

63. What is the "accounting of disclosures" requirement? A) Financial accounting of PHI-related costs B) Tracking and reporting certain PHI disclosures C) Annual reporting to OCR D) Documentation of all PHI uses

Answer: B) Tracking and reporting certain PHI disclosures. Accounting of disclosures requires covered entities to track and provide individuals with information about certain disclosures of their PHI made during the six years prior to the request.

64. When can covered entities deny amendment requests? A) When the request would be expensive to implement B) When the PHI was not created by the covered entity C) When the individual disagrees with their diagnosis D) Never - all amendment requests must be granted

Answer: B) When the PHI was not created by the covered entity. Amendment requests can be denied for several reasons including when the PHI was not created by the covered entity, is not available for access, or is accurate and complete.

65. What is required for workforce training under HIPAA? A) Annual training for all employees B) Training appropriate to workforce member's access level C) Standardized training curriculum for all staff D) Certification testing after training completion

Answer: B) Training appropriate to workforce member's access level. HIPAA requires training that is appropriate to each workforce member's level of access to PHI and their job responsibilities, rather than standardized training for all employees.

66. Which of the following is NOT a patient right under HIPAA? A) Right to access their PHI B) Right to amend their PHI C) Right to free copies of their PHI D) Right to file complaints

Answer: C) Right to free copies of their PHI. While patients have the right to access their PHI, covered entities may charge reasonable, cost-based fees for providing copies, though fees cannot be charged for electronic access through patient portals.

67. What is the purpose of the HIPAA Security Officer requirement? A) To conduct security training B) To investigate security incidents C) To develop and implement security policies D) To manage IT systems

Answer: C) To develop and implement security policies. The Security Officer is responsible for developing and implementing the covered entity's security policies and procedures, though they may delegate specific tasks to others.

68. When is individual authorization required for PHI use or disclosure? A) For all uses and disclosures B) When not permitted by

Privacy Rule exceptions C) Only for disclosures to business associates D) Only for marketing purposes

Answer: B) When not permitted by Privacy Rule exceptions. Individual authorization is required for PHI uses and disclosures unless they are specifically permitted by Privacy Rule exceptions such as treatment, payment, healthcare operations, or other specified purposes.

69. What constitutes "willful neglect" under HIPAA penalty structure? A) Accidental violations B) Violations due to lack of knowledge C) Conscious, intentional failure to comply D) Violations corrected within 30 days

Answer: C) Conscious, intentional failure to comply. Willful neglect involves conscious, intentional failure or reckless indifference to HIPAA requirements, resulting in higher penalty amounts than unintentional violations.

70. How long must HIPAA documentation be retained? A) Three years B) Five years C) Six years D) Ten years

Answer: C) Six years. HIPAA requires retention of documentation including policies, procedures, training records, and other compliance documentation for six years from the date of creation or last effective date.

71. What is required for PHI transmitted over electronic networks? A) Encryption is always required B) End-to-end encryption must be used C) Appropriate security measures based on risk assessment D) Virtual private networks must be used

Answer: C) Appropriate security measures based on risk assessment. HIPAA requires appropriate security measures for electronic PHI transmission based on risk assessment, which may include encryption but does not mandate specific technologies.

72. Which of the following breaches requires individual notification? A) All unauthorized access to PHI B) Only breaches affecting 500 or more individuals C) Breaches posing significant risk of financial harm D) Breaches not meeting regulatory exceptions

Answer: D) Breaches not meeting regulatory exceptions. Individual notification is required for all breaches unless they fall under specific regulatory exceptions, regardless of the number of individuals affected or type of harm.

73. What is the "breach notification rule" deadline for individual notification? A) 30 days after discovery B) 60 days after discovery C) 90 days after discovery D) Without unreasonable delay, no later than 60 days

Answer: D) Without unreasonable delay, no later than 60 days. Individual breach notification must be provided without unreasonable delay and no later than 60 days after discovery of the breach.

74. When can covered entities use or disclose PHI for fundraising? A) Never without individual authorization B) For any fundraising activities C) With opportunity for individuals to opt out D) Only with prior written consent

Answer: C) With opportunity for individuals to opt out. PHI may be used for fundraising activities related to the covered entity, but individuals must be provided with an opportunity to opt out of future fundraising communications.

75. What is required for workforce access controls? A) All workforce members must have identical access B) Access based on job responsibilities and minimum necessary C) Unlimited access for all clinical staff D) Access only to PHI created by the workforce member

Answer: B) Access based on job responsibilities and minimum necessary. Workforce access controls must limit PHI access to what is necessary for each workforce member's job responsibilities, implementing the minimum necessary standard.

76. Which sanctions are required for workforce violations of HIPAA policies? A) Immediate termination B) Progressive discipline starting with warnings C) Appropriate sanctions applied consistently D) Financial penalties only

Answer: C) Appropriate sanctions applied consistently. HIPAA requires appropriate sanctions for workforce members who violate policies, but does not mandate specific disciplinary measures. Sanctions should be applied consistently and proportionally.

77. What is the purpose of audit controls under the Security Rule? A) To prevent unauthorized access B) To record and examine access to ePHI C) To encrypt PHI during transmission D) To backup electronic systems

Answer: B) To record and examine access to ePHI. Audit controls implement procedures to record and examine access and other activity in information systems containing or using electronic PHI.

78. When can parents access their minor child's PHI? A) Always until the child reaches age 18 B) When authorized by state law or healthcare provider C) Never without the child's permission D) Only for medical emergencies

Answer: B) When authorized by state law or healthcare provider. Parent access to minor child PHI depends on state law and healthcare provider policies. Some situations may limit parent access, such as when minors can consent to treatment independently.

79. What constitutes "health information" under HIPAA? A) Only medical records and test results B) Any information about health status, provision of care, or payment C) Information created by healthcare providers only D) Diagnostic and treatment information only

Answer: B) Any information about health status, provision of care, or payment. Health information broadly includes any information relating to an individual's physical or mental health, provision of healthcare, or payment for healthcare services.

80. How should covered entities handle requests for PHI from law enforcement? A) Always comply immediately B) Always require a warrant C) Follow specific Privacy Rule provisions for law enforcement D) Require individual authorization in all cases

Answer: C) Follow specific Privacy Rule provisions for law enforcement. The Privacy Rule includes specific provisions governing disclosures to law enforcement, which vary depending on the purpose and circumstances of the request.

81. What is required for contingency planning under the Security Rule? A) Daily backups of all systems B) Procedures for emergency access and data recovery C) Duplicate facilities in different locations D) 24-hour system monitoring

Answer: B) Procedures for emergency access and data recovery. Contingency planning requires procedures for responding to emergencies or other occurrences that damage systems containing ePHI, including emergency access procedures and data backup and recovery plans.

82. When can PHI be used for research purposes? A) Never without individual authorization B) With IRB or Privacy Board approval under specific conditions C) Always for public health research D) Only for research conducted by the covered entity

Answer: B) With IRB or Privacy Board approval under specific conditions. PHI may be used for research with individual authorization, IRB or Privacy Board approval, or under other specific Privacy Rule provisions for research activities.

83. What is the "minimum necessary" exception for treatment activities? A) Minimum necessary always applies to treatment B) Minimum necessary never applies to treatment C) Minimum necessary applies only to referrals D) Minimum necessary depends on the type of treatment

Answer: B) Minimum necessary never applies to treatment. The minimum necessary standard does not apply to disclosures for treatment purposes, allowing healthcare providers to share whatever PHI they believe is necessary for patient care.

84. How should covered entities respond to individual complaints about HIPAA violations? A) Investigate only written complaints B) Dismiss complaints without merit C) Investigate all complaints and document responses D) Refer all complaints to legal counsel

Answer: C) Investigate all complaints and document responses. Covered entities must process complaints about their privacy practices, investigate concerns, and document their responses, regardless of how complaints are submitted.

85. What is required when PHI is disposed of or destroyed? A) Witness verification of destruction B) Reasonable safeguards to prevent unauthorized access C) Certificate of destruction from vendor D) Individual notification of disposal

Answer: B) Reasonable safeguards to prevent unauthorized access. PHI disposal must include reasonable safeguards to prevent unauthorized access to the information during the disposal process, but specific methods are not mandated.

86. Which workforce members require HIPAA training? A) Only those with direct patient contact B) Only clinical staff members C) All workforce members with PHI access D) Only newly hired employees

Answer: C) All workforce members with PHI access. All workforce members who have access to PHI must receive appropriate

HIPAA training, regardless of their role or department within the organization.

87. What is the purpose of information system activity review? A) To improve system performance B) To identify security incidents and unauthorized access C) To plan system upgrades D) To train users on system functions

Answer: B) To identify security incidents and unauthorized access. Information system activity review involves regularly reviewing system logs and access reports to identify potential security incidents and unauthorized access to ePHI.

88. When can covered entities charge fees for PHI access? A) Never B) For all copies provided to individuals C) Only for paper copies, not electronic access D) Reasonable, cost-based fees for copies

Answer: D) Reasonable, cost-based fees for copies. Covered entities may charge reasonable, cost-based fees for providing copies of PHI, but cannot charge fees for individuals' electronic access to their own information through patient portals.

89. What is required for PHI amendment procedures? A) All amendments must be approved B) Procedures for reviewing and responding to amendment requests C) Individual authorization for all amendments D) Medical staff approval for clinical amendments

Answer: B) Procedures for reviewing and responding to amendment requests. Covered entities must have procedures for reviewing amendment requests and responding to individuals, though not all amendment requests must be granted.

90. How should covered entities handle oral PHI disclosures? A) Oral disclosures are not covered by HIPAA B) Same protections as written PHI C) Oral disclosures require individual authorization D) Document all oral disclosures

Answer: B) Same protections as written PHI. HIPAA protections apply to PHI in all forms, including oral communications. The same privacy and security requirements apply regardless of the form of PHI.

91. What constitutes "administrative safeguards" under the Security Rule? A) Physical protection of computer systems B) Technology controls for system access C) Policies and procedures for managing security D) Encryption of stored data

Answer: C) Policies and procedures for managing security. Administrative safeguards consist of policies, procedures, and processes for managing the conduct of workforce members and electronic information systems containing ePHI.

92. When must business associate agreements be in place? A) Before any PHI is shared B) Within 30 days of relationship beginning C) Only for written contracts D) After business relationships are established

Answer: A) Before any PHI is shared. Business associate agreements must be executed before PHI is shared with business associates, establishing required privacy and security protections before the relationship begins.

93. What is the "psychotherapy notes" exception? A) Psychotherapy notes cannot be accessed by individuals B) Special authorization requirements for psychotherapy notes C) Psychotherapy notes are not considered PHI D) Psychotherapy notes require court orders for disclosure

Answer: B) Special authorization requirements for psychotherapy notes. Psychotherapy notes have heightened protection under HIPAA, requiring specific individual authorization for most uses and disclosures beyond treatment by the note creator.

94. How should covered entities handle PHI of deceased individuals? A) HIPAA protections end at death B) Protections continue for 50 years after death C) Protections continue indefinitely D) Only family members can access deceased PHI

Answer: B) Protections continue for 50 years after death. HIPAA protections for PHI of deceased individuals continue for 50 years after death, after which the information is no longer considered protected health information.

95. What is required for workforce termination procedures? A) Return of all PHI copies B) Termination of system access C) Exit interviews about HIPAA compliance D) Notification to business associates

Answer: B) Termination of system access. Security Rule requires procedures for terminating access to ePHI when workforce members no longer need access, particularly when employment or other relationships end.

96. Which disclosures are exempt from accounting requirements? A) Disclosures for treatment, payment, and operations B) Only disclosures with individual authorization C) Disclosures to family members D) All disclosures must be included in accounting

Answer: A) Disclosures for treatment, payment, and operations. Accounting of disclosures does not include routine disclosures for treatment, payment, healthcare operations, individual authorization, or several other specified categories.

97. What is required for unique user identification? A) All users must have different passwords B) Each user must have a unique identifier for system access C) Biometric authentication for all users D) Two-factor authentication requirements

Answer: B) Each user must have a unique identifier for system access. Unique user identification requires assigning each system user a unique name or number that identifies and tracks their identity and access to ePHI.

98. How should covered entities handle marketing communications? A) Marketing requires individual authorization B) All healthcare communications are permitted C) Marketing is prohibited under HIPAA D) Marketing requires opt-out opportunities only

Answer: A) Marketing requires individual authorization. Most marketing communications require individual authorization under HIPAA, though some healthcare communications and treatment alternatives may be permitted without authorization.

99. What is the "facility access controls" requirement? A) Badge access systems for all employees B) Procedures to limit physical access to ePHI systems C) Security guards at all entrances D) Visitor logs for all facilities

Answer: B) Procedures to limit physical access to ePHI systems. Facility access controls require procedures to limit physical access to electronic information systems and equipment housing ePHI while ensuring appropriate access for authorized personnel.

100. When can covered entities disclose PHI for public health activities? A) Never without individual authorization B) For any public health purpose C) As permitted by Privacy Rule public health provisions D) Only during declared health emergencies

Answer: C) As permitted by Privacy Rule public health provisions. PHI may be disclosed for public health activities as specifically permitted by Privacy Rule provisions, which include reporting to public health authorities, preventing disease spread, and other specified public health purposes.

Fraud and Abuse Laws

101. What does the Stark Law primarily prohibit? A) Kickbacks to healthcare providers B) Physician self-referrals for designated health services C) False claims to government programs D) Unlicensed practice of medicine

Answer: B) Physician self-referrals for designated health services. The Stark Law prohibits physicians from referring patients for designated health services to entities with which they have financial relationships, unless a specific exception applies.

102. How many designated health services are covered under the Stark Law? A) 9 categories B) 10 categories C) 11 categories D) 12 categories

Answer: C) 11 categories. The Stark Law covers 11 categories of designated health services including clinical laboratory services, physical therapy, radiology, radiation therapy, durable medical equipment, parenteral and enteral nutrients, prosthetics, home health services, outpatient prescription drugs, and inpatient and outpatient hospital services.

103. What type of liability does the Stark Law impose? A) Criminal liability only B) Civil liability only C) Strict liability D) Intent-based liability

Answer: C) Strict liability. The Stark Law imposes strict liability, meaning violations can occur without proof of intent to violate the law. Technical non-compliance can result in penalties even without fraudulent intent.

104. What is the "knows or should know" standard under Stark Law? A) Actual knowledge of violations is required B) Constructive knowledge based on available information C) Intent to violate the law must be proven D) Reasonable belief that conduct is legal

Answer: B) Constructive knowledge based on available information. The "knows or should know" standard means that actual knowledge is not required; constructive knowledge based on information that is reasonably available can establish liability.

105. Which of the following is NOT a Stark Law exception? A) Employment relationships B) Personal service arrangements C) Fair market value transactions D) Safe harbor compliance

Answer: D) Safe harbor compliance. Safe harbors apply to the Anti-Kickback Statute, not the Stark Law. Stark Law has its own exceptions including employment, personal services, and fair market value arrangements.

106. What is required for the Anti-Kickback Statute to apply? A) Services must be provided to Medicare patients only B) Services must be payable by any federal healthcare program C) Intent to induce referrals must be proven D) Both B and C

Answer: D) Both B and C. The Anti-Kickback Statute applies to services payable by federal healthcare programs and requires proof of intent to induce referrals, unlike Stark Law's strict liability standard.

107. What is the "one purpose" test under the Anti-Kickback Statute? A) Arrangements must have only one business purpose B) If one purpose is to induce referrals, AKS may be violated C) Only the primary purpose of arrangements matters D) All purposes must be legitimate to avoid violations

Answer: B) If one purpose is to induce referrals, AKS may be violated. Under the "one purpose" test, if even one purpose of an arrangement is to induce referrals, the Anti-Kickback Statute may be violated, even if other legitimate purposes exist.

108. How do Anti-Kickback Statute safe harbors differ from Stark Law exceptions? A) Safe harbors are mandatory while exceptions are optional B) Safe harbors provide complete protection while exceptions are required C) There is no difference between safe harbors and exceptions D) Safe harbors apply only to criminal enforcement

Answer: B) Safe harbors provide complete protection while exceptions are required. AKS safe harbors provide complete protection from prosecution but are optional, while Stark Law exceptions are required for arrangements to be permissible - there is no partial compliance protection.

109. What is the relationship between the Anti-Kickback Statute and False Claims Act? A) They are completely separate laws with no relationship B) AKS violations automatically create False Claims Act violations C) Claims resulting from AKS violations are false claims D) False Claims Act supersedes Anti-Kickback Statute

Answer: C) Claims resulting from AKS violations are false claims. The Affordable Care Act established that claims resulting from AKS violations are automatically considered false claims for civil enforcement purposes, linking the two statutes.

110. What does "remuneration" include under the Anti-Kickback Statute? A) Only cash payments B) Anything of value regardless of form C) Only financial benefits above $100 D) Only direct payments to physicians

Answer: B) Anything of value regardless of form. Remuneration under AKS includes anything of value transferred between parties, regardless of form, amount, or characterization, including cash, gifts, services, or other benefits.

111. What is required to prove a False Claims Act violation? A) Criminal intent to defraud the government B) Actual financial harm to government programs C) Knowingly submitting false or fraudulent claims D) Pattern of violations over extended time

Answer: C) Knowingly submitting false or fraudulent claims. FCA violations require proof that false or fraudulent claims were submitted "knowingly," which includes actual knowledge, deliberate ignorance, or reckless disregard for truth or falsity.

112. What does "knowingly" mean under the False Claims Act? A) Actual knowledge of falsity only B) Actual knowledge, deliberate ignorance, or reckless disregard C) Intent to defraud the government D) Negligent submission of incorrect claims

Answer: B) Actual knowledge, deliberate ignorance, or reckless disregard. The FCA's "knowingly" standard includes actual knowledge of falsity, deliberate ignorance of truth or falsity, or reckless disregard for the truth or falsity of information.

113. What are the potential penalties under the False Claims Act? A) Criminal fines only B) Civil monetary penalties only C) Treble damages plus civil monetary penalties D) Exclusion from federal programs only

Answer: C) Treble damages plus civil monetary penalties. FCA penalties include treble (three times) damages equal to the government's actual damages plus civil monetary penalties for each false claim submitted.

114. What is a "qui tam" lawsuit? A) Government enforcement action B) Private party lawsuit on behalf of the government C) Class action lawsuit by patients D) Regulatory investigation by OIG

Answer: B) Private party lawsuit on behalf of the government. Qui tam lawsuits are filed by private parties (relators or whistleblowers) on behalf of the government, allowing them to share in any recovery if the lawsuit is successful.

115. What protection do qui tam relators receive? A) Financial rewards only B) Anonymity throughout the process C) Protection from retaliation D) Immunity from prosecution

Answer: C) Protection from retaliation. Qui tam relators receive protection from retaliation by their employers for reporting suspected violations or participating in investigations or enforcement actions.

116. What is "fair market value" under healthcare compliance laws? A) The highest price paid for similar services B) Compensation that would be paid in arm's length transactions C) Government-established rates for healthcare services D) Average compensation in the local market

Answer: B) Compensation that would be paid in arm's length transactions. Fair market value is compensation that would be paid in arm's length transactions for identical or similar services in the same geographic area, without considering referral volume or value.

117. Which of the following creates the highest risk for Stark Law violations? A) Physician employment agreements B) Medical director agreements C) Equipment lease arrangements D) All have equal risk if not properly structured

Answer: D) All have equal risk if not properly structured. All physician financial arrangements create Stark Law risk if not properly structured to meet applicable exceptions. The risk level depends on compliance with exception requirements rather than arrangement type.

118. What is the "volume or value" test under Stark Law? A) Arrangements must involve minimum dollar amounts B) Compensation cannot take into account referral volume or value C) Large arrangements receive greater scrutiny D) High-volume referrals are automatically prohibited

Answer: B) Compensation cannot take into account referral volume or value. Most Stark Law exceptions prohibit compensation that takes into account the volume or value of referrals, ensuring that financial relationships don't create incentives for inappropriate referrals.

119. What is required for Stark Law employment exception compliance? A) Full-time employment only B) Bona fide employment with fair market value compensation C) Employment contracts of at least one year D) Board certification in relevant specialty

Answer: B) Bona fide employment with fair market value compensation. The employment exception requires bona fide employment relationships with compensation that is fair market value, commercially reasonable, and not based on referral volume or value.

120. How do value-based arrangements affect Stark Law compliance? A) They are automatically exempt from Stark Law B) New exceptions provide flexibility for value-based care C) They increase Stark Law compliance risks D) They are prohibited under current Stark Law

Answer: B) New exceptions provide flexibility for value-based care. Recent Stark Law changes introduced new exceptions for value-based arrangements that allow greater flexibility for financial relationships focused on coordinating care and improving outcomes.

121. What is the government's burden of proof in False Claims Act cases? A) Beyond a reasonable doubt B) Clear and convincing evidence C) Preponderance of the evidence D) Probable cause

Answer: C) Preponderance of the evidence. FCA civil cases require proof by a preponderance of the evidence, a lower standard than the "beyond a reasonable doubt" required for criminal cases.

122. Which federal agency primarily enforces the Anti-Kickback Statute? A) Centers for Medicare & Medicaid Services B) Office of Inspector General C) Department of Justice D) Food and Drug Administration

Answer: B) Office of Inspector General. The Office of Inspector General has primary responsibility for Anti-Kickback Statute enforcement, though the Department of Justice prosecutes criminal and civil cases.

123. What is "commercial reasonableness" under Stark Law exceptions? A) Compensation at market rates B) Arrangements that make business sense apart from referrals C) Profitable arrangements for both parties D) Arrangements approved by compliance officers

Answer: B) Arrangements that make business sense apart from referrals. Commercial reasonableness means arrangements would make business sense and be entered into even if no referrals were made between the parties.

124. When might self-disclosure to the government be appropriate? A) For all potential violations B) Only for criminal violations C) When violations are significant and corrective action is needed D) Never - organizations should wait for government investigation

Answer: C) When violations are significant and corrective action is needed. Self-disclosure may be appropriate for significant violations where organizations want to demonstrate cooperation and good faith, though each situation requires careful analysis.

125. What is the statute of limitations for False Claims Act violations? A) Three years from discovery B) Six years from violation or three years from discovery C) Five years from violation D) No statute of limitations

Answer: B) Six years from violation or three years from discovery. FCA has a six-year statute of limitations from violation date or three years from when material facts were known or should have been known by government officials, whichever is later.

126. Which arrangement requires the most careful Stark Law analysis? A) Hospital-physician employment B) Medical staff membership C) Emergency call coverage D) Physician investment in hospital

Answer: D) Physician investment in hospital. Physician investment in hospitals creates ownership interests that are particularly complex under Stark Law and often require the most careful analysis to ensure compliance.

127. What is the "group practice" exception under Stark Law? A) All physicians in a practice can refer to each other freely B) Specific requirements for physician groups to refer within the group C) Groups of hospitals can share physician services D) Refers to academic medical center arrangements

Answer: B) Specific requirements for physician groups to refer within the group. The group practice exception allows physicians in qualifying groups to refer within their group for designated health services, but the group must meet specific organizational and financial requirements.

128. How do state laws interact with federal fraud and abuse laws? A) State laws supersede federal laws B) Federal laws supersede all state laws C) Both federal and state laws may apply D) Only federal laws apply to healthcare

Answer: C) Both federal and state laws may apply. Healthcare organizations must comply with both federal fraud and abuse laws and applicable state laws, which may have different or additional requirements.

129. What is the "indirect compensation arrangement" concept? A) Payments made through third parties B) Compensation arrangements involving multiple parties C) Non-cash compensation between parties D) Arrangements where compensation flows indirectly between referring physicians and DHS entities

Answer: D) Arrangements where compensation flows indirectly between referring physicians and DHS entities. Indirect compensation arrangements involve unbroken chains of financial relationships between referring physicians and DHS entities, even when they don't have direct financial relationships.

130. Which safe harbor protects legitimate space rental arrangements? A) Employment safe harbor B) Personal services safe harbor C) Space rental safe harbor D) Fair market value safe harbor

Answer: C) Space rental safe harbor. The space rental safe harbor protects legitimate real estate leasing arrangements that meet specific requirements including written agreements, fair market value rent, and exclusive use during rental periods.

131. What documentation is most important for Stark Law compliance? A) Physician referral logs B) Financial arrangement contracts and fair market value assessments C) Patient outcome data D) Marketing materials and communications

Answer: B) Financial arrangement contracts and fair market value assessments. Proper documentation of financial arrangements, including written contracts and fair market value assessments, is essential for demonstrating Stark Law exception compliance.

132. How often should physician financial arrangements be reviewed for compliance? A) Monthly B) Quarterly C) Annually D) Only when problems are identified

Answer: C) Annually. Most healthcare organizations review physician financial arrangements annually to ensure continued compliance with applicable exceptions and fair market value requirements.

133. What is the "whole hospital" exception under Stark Law? A) Hospitals can refer to any affiliated entity B) Physician investment in entire hospitals may be permitted C) Hospital systems are exempt from Stark Law D) All hospital services are designated health services

Answer: B) Physician investment in entire hospitals may be permitted. The whole hospital exception allows physician investment in entire hospitals under specific circumstances, though it has strict requirements and limited applicability.

134. Which factor is most important in Anti-Kickback Statute enforcement? A) Dollar amount of remuneration B) Number of referrals generated C) Intent to induce referrals D) Type of healthcare service

Answer: C) Intent to induce referrals. Intent to induce referrals is the key element for AKS violations, as the statute requires proof of intent, unlike Stark Law's strict liability standard.

135. What is "upcoding" in healthcare billing? A) Using new procedure codes B) Billing for higher-paying services than provided C) Correcting billing errors after submission D) Adding modifier codes to claims

Answer: B) Billing for higher-paying services than provided. Upcoding involves billing for higher-level or more expensive services than were actually provided or documented, creating false claims liability.

136. How do Corporate Integrity Agreements relate to fraud and abuse laws? A) They prevent fraud and abuse violations B) They are settlement agreements that include ongoing compliance obligations C) They replace criminal prosecution D) They apply only to physician practices

Answer: B) They are settlement agreements that include ongoing compliance obligations. Corporate Integrity Agreements are settlement agreements between healthcare organizations and OIG that establish ongoing compliance monitoring and reporting requirements for specified periods.

137. What is the "anti-markup" rule? A) Prohibition on charging above cost for services B) Limits on billing markups for certain laboratory services C) Requirements for transparent pricing D) Restrictions on pharmaceutical pricing

Answer: B) Limits on billing markups for certain laboratory services. The anti-markup rule limits how much physicians can mark up certain laboratory services when billing Medicare, preventing excessive profits on referral services.

138. Which violation typically results in the highest penalties? A) Technical Stark Law violations B) Intent-based Anti-Kickback violations C) False Claims Act violations with large dollar amounts D) All violations have equal penalties

Answer: C) False Claims Act violations with large dollar amounts. FCA violations often result in the highest penalties due to treble damages and per-claim penalties, particularly when involving large numbers of claims or high dollar amounts.

139. What is required for "medical necessity" in healthcare billing? A) Physician determination alone is sufficient B) Services must be reasonable and necessary for diagnosis or treatment C) Prior authorization from insurance companies D) Documentation of patient request for services

Answer: B) Services must be reasonable and necessary for diagnosis or treatment. Medical necessity requires that services be reasonable and necessary for diagnosis or treatment of patient conditions, with adequate documentation supporting this determination.

140. How do fraud and abuse laws apply to telehealth services? A) Telehealth is exempt from fraud and abuse laws B) Special rules apply only to telehealth C) Same laws apply with potential enforcement challenges D) Only state laws regulate telehealth fraud

Answer: C) Same laws apply with potential enforcement challenges. Fraud and abuse laws apply to telehealth services just as they do to in-person services, though enforcement may face additional challenges related to cross-state practice and documentation.

141. What is the "beneficiary inducement" prohibition? A) Preventing patients from choosing providers B) Prohibiting inducements to Medicare/Medicaid beneficiaries C) Requiring patient copayments D) Limiting marketing to patients

Answer: B) Prohibiting inducements to Medicare/Medicaid beneficiaries. Beneficiary inducement laws prohibit offering inducements to Medicare and Medicaid beneficiaries to receive services, with exceptions for certain nominal value gifts and services.

142. Which arrangement creates the lowest fraud and abuse risk? A) Physician recruitment agreements B) Medical directorships C) Fair market value consulting agreements D) None - all arrangements have equal risk

Answer: D) None - all arrangements have equal risk. All physician financial arrangements create fraud and abuse risk that must be managed through proper structuring, documentation, and compliance with applicable exceptions and safe harbors.

143. What is the "incident to" billing concept? A) Billing for services provided during emergencies B) Billing physician services for services provided by non-physicians C) Billing for services provided at multiple locations D) Billing for services provided to hospital inpatients

Answer: B) Billing physician services for services provided by non-physicians. "Incident to" billing allows physician billing for certain services provided by non-physician practitioners under physician supervision, with specific requirements for supervision and billing.

144. How do quality incentive payments affect fraud and abuse compliance? A) They are automatically exempt from fraud and abuse laws B) They may create compliance risks requiring careful structuring C) They are prohibited under current fraud and abuse laws D) They apply only to government payment programs

Answer: B) They may create compliance risks requiring careful structuring. Quality incentive payments can create fraud and abuse compliance risks if not properly structured to ensure payments are based on actual quality achievements rather than referral volume.

145. What is the most effective approach for fraud and abuse compliance? A) Avoiding all physician financial relationships B) Obtaining legal opinions for all arrangements C) Systematic risk assessment and mitigation D) Relying on safe harbor and exception compliance only

Answer: C) Systematic risk assessment and mitigation. Effective fraud and abuse compliance requires systematic risk assessment, appropriate arrangement structuring, ongoing monitoring, and comprehensive compliance programs rather than avoidance or minimal compliance approaches.

146. Which documentation is most important for False Claims Act defense? A) Physician credentials and licenses B) Medical records supporting billed services C) Insurance authorization letters D) Patient satisfaction surveys

Answer: B) Medical records supporting billed services. Medical records that adequately document and support billed services provide the best defense against False Claims Act allegations by demonstrating that claims were accurate and appropriate.

147. What is the "signature on file" requirement? A) Electronic signatures are prohibited B) Original signatures must be maintained for all orders C) Physicians must personally sign all documentation D) Adequate documentation of physician orders and authentication

Answer: D) Adequate documentation of physician orders and authentication. Signature on file requirements ensure that physician orders are properly documented and authenticated, with various methods acceptable for demonstrating physician involvement in patient care decisions.

148. How do fraud and abuse laws apply to pharmaceutical relationships? A) They don't apply to pharmaceutical companies B) Special exemptions exist for drug companies C) Same principles apply with industry-specific considerations D) Only criminal laws apply to pharmaceutical relationships

Answer: C) Same principles apply with industry-specific considerations. Fraud and abuse laws apply to pharmaceutical company relationships with healthcare providers, with industry-specific considerations for research, education, and promotional activities.

149. What is the "Sunshine Act" reporting requirement? A) Public reporting of hospital quality measures B) Transparency reporting of pharmaceutical and device company payments C) Financial disclosure by healthcare executives D) Public reporting of healthcare pricing

Answer: B) Transparency reporting of pharmaceutical and device company payments. The Physician Payments Sunshine Act requires pharmaceutical and device companies to report payments and transfers of value to physicians and teaching hospitals for public transparency.

150. Which trend is most important for future fraud and abuse enforcement? A) Decreased government enforcement activity B)

Focus on intent-based violations only C) Increased emphasis on value-based care arrangements D) Elimination of safe harbors and exceptions

Answer: C) Increased emphasis on value-based care arrangements. Future enforcement is likely to focus increasingly on value-based care arrangements as healthcare moves away from fee-for-service toward payment models that reward quality and outcomes.

Regulatory Requirements and Quality Programs

151. What does CMS stand for? A) Center for Medical Services B) Centers for Medicare & Medicaid Services C) Centers for Medical Safety D) Committee for Medical Standards

Answer: B) Centers for Medicare & Medicaid Services. CMS is the Centers for Medicare & Medicaid Services, the federal agency that administers Medicare, Medicaid, and other federal healthcare programs.

152. How many Conditions of Participation categories apply to hospitals? A) 12 categories B) 16 categories C) 20 categories D) 24 categories

Answer: B) 16 categories. Hospital Conditions of Participation include 16 categories covering areas such as governing body, medical staff, nursing services, patient rights, quality assurance, and infection control.

153. What is the maximum penalty for termination from Medicare participation? A) Financial penalties only B) Loss of all

Medicare reimbursement C) Probationary status with monitoring D) Temporary suspension of new admissions

Answer: B) Loss of all Medicare reimbursement. Termination from Medicare participation results in complete loss of Medicare reimbursement, which can threaten organizational survival for most healthcare providers.

154. What triggers immediate jeopardy findings under CMS regulations? A) Any condition-level deficiency B) Situations posing immediate threat to patient health and safety C) Failure to meet documentation requirements D) Staffing levels below minimum requirements

Answer: B) Situations posing immediate threat to patient health and safety. Immediate jeopardy findings occur when deficiencies are identified that pose immediate threat to patient health and safety, requiring immediate correction to avoid termination.

155. How often must hospitals conduct community health needs assessments? A) Annually B) Every two years C) Every three years D) Every five years

Answer: C) Every three years. Nonprofit hospitals must conduct community health needs assessments every three years and develop implementation strategies to address identified health needs.

156. What is the 60-day overpayment rule? A) Claims must be paid within 60 days B) Overpayments must be identified within 60 days C) Overpayments must be returned within 60 days of identification D) Appeals must be filed within 60 days

Answer: C) Overpayments must be returned within 60 days of identification. The 60-day rule requires healthcare organizations to return overpayments within 60 days of identification, with failure to comply potentially creating False Claims Act liability.

157. What does MIPS stand for in CMS quality programs? A) Medicare Incentive Payment System B) Merit-based Incentive Payment System C) Medical Improvement Payment System D) Medicare Integration Payment System

Answer: B) Merit-based Incentive Payment System. MIPS is the Merit-based Incentive Payment System that adjusts Medicare payments based on physician performance in quality, cost, improvement activities, and promoting interoperability.

158. How many performance categories are included in MIPS scoring? A) Three categories B) Four categories C) Five categories D) Six categories

Answer: B) Four categories. MIPS includes four performance categories: Quality, Cost, Improvement Activities, and Promoting Interoperability, each contributing to the overall composite performance score.

159. What is the minimum MIPS composite score to avoid payment penalties? A) 30 points B) 45 points C) 60 points D) 75 points

Answer: D) 75 points. The MIPS performance threshold varies annually but typically requires scores above 75 points to avoid payment penalties, with higher scores earning bonus payments.

160. What does ACO stand for in value-based care? A) Accountable Care Organization B) Administrative Care Office C) Advanced Care Option D) Associated Care Organization

Answer: A) Accountable Care Organization. ACO stands for Accountable Care Organization, a group of healthcare providers that coordinate care for defined patient populations and share financial responsibility for outcomes.

161. What is required for Medicare Shared Savings Program participation? A) Hospital ownership of physician practices B) Minimum of 5,000 assigned beneficiaries C) Acceptance of downside financial risk D) Electronic health record implementation

Answer: B) Minimum of 5,000 assigned beneficiaries. Medicare Shared Savings Program ACOs must have sufficient numbers of assigned beneficiaries to enable meaningful measurement of quality and cost performance, typically at least 5,000 beneficiaries.

162. How many quality measures are included in MSSP ACO reporting? A) 25 measures B) 30 measures C) 33 measures D) 40 measures

Answer: C) 33 measures. Medicare Shared Savings Program ACOs report on 33 quality measures addressing patient safety, clinical care, patient experience, and population health domains.

163. What does eCQM stand for in quality reporting? A) Electronic Clinical Quality Measures B) Electronic Care Quality Management C) Enhanced Clinical Quality Metrics D) Electronic Compliance Quality Measures

Answer: A) Electronic Clinical Quality Measures. eCQMs are Electronic Clinical Quality Measures that use electronic health record data to calculate quality performance automatically rather than through manual chart abstraction.

164. What is the primary advantage of eCQMs over traditional quality measures? A) Lower implementation costs B) Reduced administrative burden and improved accuracy C) Better patient outcomes D) Increased physician satisfaction

Answer: B) Reduced administrative burden and improved accuracy. eCQMs reduce administrative burden by automating data collection and calculation while improving accuracy by eliminating manual chart abstraction errors.

165. What does EMTALA require for emergency department patients? A) Insurance verification before treatment B) Medical screening examination for all patients C) Physician evaluation within 30 minutes D) Advance payment for services

Answer: B) Medical screening examination for all patients. EMTALA requires medical screening examinations for all individuals seeking emergency care, regardless of ability to pay or insurance status.

166. What constitutes an "emergency medical condition" under EMTALA? A) Only life-threatening conditions B) Conditions requiring immediate medical attention to prevent serious jeopardy C) Any condition requiring hospital admission D) Conditions specifically listed in EMTALA regulations

Answer: B) Conditions requiring immediate medical attention to prevent serious jeopardy. EMTALA defines emergency medical conditions broadly as those requiring immediate medical attention to prevent serious jeopardy to patient health, serious impairment to bodily functions, or serious dysfunction of body organs.

167. What is required before transferring EMTALA patients? A) Patient consent only B) Insurance authorization C) Stabilization or appropriate transfer procedures D) Physician consultation

Answer: C) Stabilization or appropriate transfer procedures. EMTALA requires that emergency medical conditions be stabilized before transfer, or if stabilization is not possible, that appropriate transfer procedures be followed including physician certification and receiving hospital acceptance.

168. What does "meaningful use" refer to in healthcare? A) Appropriate utilization of healthcare services B) Electronic health record implementation and use requirements C) Cost-effective treatment protocols D) Patient satisfaction measurement

Answer: B) Electronic health record implementation and use requirements. Meaningful use refers to requirements for electronic health record implementation and use that demonstrate providers are using EHR technology to improve patient care quality and safety.

169. What is the current name for the meaningful use program? A) EHR Incentive Program B) Health Information Technology Program C) Promoting Interoperability Program D) Digital Health Initiative

Answer: C) Promoting Interoperability Program. The meaningful use program was renamed the Promoting Interoperability Program to emphasize data sharing and care coordination rather than just EHR adoption.

170. What does information blocking prohibit? A) Sharing patient information without consent B) Practices that interfere with access, exchange, or use of electronic health information C) Using electronic health records for billing purposes D) Transmitting health information over networks

Answer: B) Practices that interfere with access, exchange, or use of electronic health information. Information blocking regulations prohibit practices that are likely to interfere with the access, exchange, or use of electronic health information, promoting healthcare data interoperability.

171. How many information blocking exceptions exist? A) Six exceptions B) Seven exceptions C) Eight exceptions D) Ten exceptions

Answer: C) Eight exceptions. Information blocking regulations include eight exceptions that allow certain practices when they serve legitimate purposes such as privacy protection, security, safety, health IT performance, content and manner, fees, licensing, and infeasibility.

172. What is required for patient access to electronic health information? A) Patients must request access in writing B) Providers must give patients electronic access without special effort C) Access can be denied if patients cannot pay fees D) Only summary information must be provided

Answer: B) Providers must give patients electronic access without special effort. Information blocking regulations require that patients be given access to their electronic health information without special effort, typically through patient portals or similar electronic means.

173. What does FQHC stand for in healthcare? A) Federal Quality Healthcare Center B) Federally Qualified Health Center C) Federal Qualified Healthcare Clinic D) Federal Quality Health Certification

Answer: B) Federally Qualified Health Center. FQHC stands for Federally Qualified Health Center, which are community-based healthcare providers that receive federal funding to provide primary care services in underserved areas.

174. What is required for FQHC designation? A) Hospital affiliation B) Nonprofit status only C) Service to medically underserved populations with specific requirements D) Physician ownership

Answer: C) Service to medically underserved populations with specific requirements. FQHC designation requires serving medically underserved populations while meeting specific requirements for governance, services provided, and patient access regardless of ability to pay.

175. What does RHC stand for in rural healthcare? A) Rural Health Center B) Rural Healthcare Clinic C) Regional Health Center D) Rural Hospital Corporation

Answer: B) Rural Healthcare Clinic. RHC stands for Rural Health Clinic, which are healthcare facilities located in rural areas that provide primary care services and receive special Medicare and Medicaid reimbursement.

176. What is the minimum distance requirement for Critical Access Hospital designation? A) 15 miles from another hospital B) 25 miles from another hospital C) 35 miles from another hospital D) 50 miles from another hospital

Answer: C) 35 miles from another hospital. Critical Access Hospitals must generally be located more than 35 miles from another hospital (or 15 miles in mountainous terrain) to qualify for the special designation and reimbursement.

177. What is the maximum bed capacity for Critical Access Hospitals? A) 15 beds B) 20 beds C) 25 beds D) 30 beds

Answer: C) 25 beds. Critical Access Hospitals are limited to 25 inpatient beds, with flexibility to use beds for acute care or swing-bed services depending on community needs.

178. What does LTCH stand for in Medicare payment systems? A) Long-Term Care Hospital B) Limited Treatment Care Hospital C) Long-Term Chronic Hospital D) Licensed Treatment Care Hospital

Answer: A) Long-Term Care Hospital. LTCH stands for Long-Term Care Hospital, which are facilities that provide extended medical and rehabilitative care for patients requiring longer hospital stays.

179. What is the minimum average length of stay for LTCH classification? A) 20 days B) 25 days C) 30 days D) 35 days

Answer: B) 25 days. Long-Term Care Hospitals must maintain an average length of stay of greater than 25 days for Medicare patients to qualify for LTCH classification and payment rates.

180. What does SNF stand for in Medicare terminology? A) Special Nursing Facility B) Skilled Nursing Facility C) Senior Nursing Facility D) Specialized Nursing Facility

Answer: B) Skilled Nursing Facility. SNF stands for Skilled Nursing Facility, which provides skilled nursing care and rehabilitation services to patients who no longer require acute hospital care but need more than custodial care.

181. What is required for Medicare SNF coverage? A) Physician referral only B) Prior hospitalization and skilled care needs C) Insurance pre-authorization D) Family request for placement

Answer: B) Prior hospitalization and skilled care needs. Medicare SNF coverage requires a qualifying hospital stay of at least three days and demonstration that the patient needs skilled nursing care or rehabilitation services.

182. What does IRF stand for in rehabilitation services? A) Intensive Rehabilitation Facility B) Inpatient Rehabilitation Facility C) Integrated Rehabilitation Facility D) Independent Rehabilitation Facility

Answer: B) Inpatient Rehabilitation Facility. IRF stands for Inpatient Rehabilitation Facility, which provides intensive rehabilitation services for patients recovering from injuries, illnesses, or surgeries.

183. What is the "60 percent rule" for IRFs? A) 60% of patients must achieve functional improvement B) 60% of patients must have qualifying conditions C) 60% occupancy rate must be maintained D) 60% of care must be provided by licensed therapists

Answer: B) 60% of patients must have qualifying conditions. The 60 percent rule requires that at least 60% of IRF patients have one of 13 specified medical conditions that typically require intensive rehabilitation services.

184. What does IPPS stand for in Medicare payment? A) Inpatient Prospective Payment System B) Integrated Patient Payment System C) Inpatient Provider Payment System D) Individual Patient Payment System

Answer: A) Inpatient Prospective Payment System. IPPS stands for Inpatient Prospective Payment System, which pays hospitals predetermined amounts based on diagnosis-related groups (DRGs) rather than actual costs incurred.

185. What does DRG stand for in hospital payment? A) Diagnosis Review Group B) Diagnosis-Related Group C) Disease Resource Group D) Diagnostic Reimbursement Group

Answer: B) Diagnosis-Related Group. DRG stands for Diagnosis-Related Group, which are categories used to classify hospital inpatients with similar clinical conditions and resource consumption for payment purposes.

186. What does OPPS stand for in Medicare payment? A) Outpatient Prospective Payment System B) Outpatient Provider Payment System C) Outpatient Performance Payment System D) Outpatient Patient Payment System

Answer: A) Outpatient Prospective Payment System. OPPS stands for Outpatient Prospective Payment System, which pays hospitals and other providers predetermined amounts for outpatient services based on ambulatory payment classifications.

187. What does APC stand for in outpatient payment? A) Ambulatory Payment Classification B) Ambulatory Patient Category C) Ambulatory Procedure Code D) Ambulatory Payment Code

Answer: A) Ambulatory Payment Classification. APC stands for Ambulatory Payment Classification, which groups outpatient services with similar clinical characteristics and resource requirements for payment under the OPPS.

188. What is required for observation services billing? A) Minimum 8-hour stay B) Physician orders and appropriate medical necessity C) Prior authorization from Medicare D) Separate facility registration

Answer: B) Physician orders and appropriate medical necessity. Observation services require physician orders and documentation of medical necessity for extended monitoring that doesn't meet inpatient admission criteria but exceeds typical outpatient care.

189. What does HCPCS stand for in medical coding? A) Healthcare Common Procedure Coding System B) Hospital Common Procedure Classification System C) Healthcare Clinical Procedure Coding System D) Health Care Provider Coding System

Answer: A) Healthcare Common Procedure Coding System. HCPCS stands for Healthcare Common Procedure Coding System, which includes CPT codes (Level I) and additional codes for supplies, equipment, and services not covered by CPT (Level II).

190. What is the difference between CPT and HCPCS Level II codes? A) CPT codes are for physicians, HCPCS for hospitals B) CPT codes are numeric, HCPCS are alphanumeric C) CPT covers physician services, HCPCS Level II covers supplies and equipment D) There is no difference between the coding systems

Answer: C) CPT covers physician services, HCPCS Level II covers supplies and equipment. CPT codes primarily describe physician services and procedures, while HCPCS Level II codes (starting with letters) describe supplies, equipment, and services not covered by CPT codes.

191. What does LCD stand for in Medicare coverage? A) Local Coverage Decision B) Limited Coverage Determination C) Local Clinical Decision D) Licensed Coverage Document

Answer: A) Local Coverage Decision. LCD stands for Local Coverage Decision, which are Medicare Administrative Contractor determinations about whether specific services are covered under Medicare in their geographic areas.

192. What does NCD stand for in Medicare coverage? A) National Coverage Decision B) National Clinical Determination C) National Coverage Document D) National Care Decision

Answer: A) National Coverage Decision. NCD stands for National Coverage Decision, which are nationwide determinations by CMS about whether specific services or items are covered under Medicare.

193. What is required for Medicare Advantage plan compliance? A) Hospital ownership only B) CMS certification and ongoing oversight requirements C) State insurance licensing only D) Physician network contracts only

Answer: B) CMS certification and ongoing oversight requirements. Medicare Advantage plans must obtain CMS certification and meet ongoing requirements for network adequacy, quality performance, compliance programs, and beneficiary protections.

194. What does Star Ratings measure for Medicare Advantage plans? A) Financial performance only B) Quality and performance across multiple domains C) Provider satisfaction scores D) Enrollment growth rates

Answer: B) Quality and performance across multiple domains. Medicare Advantage Star Ratings evaluate plan performance across multiple domains including clinical quality, member experience, access to care, and customer service.

195. What is the maximum Star Rating for Medicare Advantage plans? A) 4 stars B) 5 stars C) 6 stars D) 10 stars

Answer: B) 5 stars. Medicare Advantage plans are rated on a 5-star scale, with 5 stars representing the highest quality performance and 1 star representing the lowest quality performance.

196. What does HEDIS stand for in quality measurement? A) Healthcare Effectiveness Data and Information Set B) Health Education Data Information System C) Healthcare Evaluation and Data Integration System D) Health Enhancement Data and Information Standards

Answer: A) Healthcare Effectiveness Data and Information Set. HEDIS stands for Healthcare Effectiveness Data and Information Set, which are standardized performance measures used to evaluate health plan quality and performance.

197. What organization maintains HEDIS measures? A) Centers for Medicare & Medicaid Services B) National Committee for Quality Assurance C) Joint Commission D) American Medical Association

Answer: B) National Committee for Quality Assurance. The National Committee for Quality Assurance (NCQA) maintains and updates HEDIS measures, which are widely used by health plans and healthcare organizations for quality measurement.

198. What does CAHPS stand for in patient experience measurement? A) Consumer Assessment of Healthcare Providers and Systems B) Clinical Assessment of Healthcare Performance and Standards C) Comprehensive Assessment of Healthcare Practices and Services D) Consumer Analysis of Healthcare Plans and Services

Answer: A) Consumer Assessment of Healthcare Providers and Systems. CAHPS stands for Consumer Assessment of Healthcare Providers and Systems, which are standardized surveys that measure patient experience with healthcare providers and health plans.

199. What is required for Joint Commission accreditation? A) Hospital ownership by nonprofit organizations B) Compliance with

Joint Commission standards and survey process C) Minimum bed capacity requirements D) Academic medical center affiliation

Answer: B) Compliance with Joint Commission standards and survey process. Joint Commission accreditation requires compliance with established standards addressing patient safety, quality of care, and organizational performance, verified through on-site surveys.

200. How often do Joint Commission accreditation surveys typically occur? A) Annually B) Every 2 years C) Every 3 years D) Every 5 years

Answer: C) Every 3 years. Joint Commission accreditation surveys typically occur every three years, with possible interim surveys or focused visits if significant compliance issues are identified.

Risk Management and Auditing

201. What is the primary purpose of healthcare compliance auditing? A) Finding violations to report to regulators B) Early detection and prevention of compliance problems C) Meeting external audit requirements D) Documenting compliance activities

Answer: B) Early detection and prevention of compliance problems. The primary purpose of compliance auditing is early detection and prevention of problems before they become serious violations, enabling proactive correction rather than reactive response.

202. Which audit approach examines activities before they occur?
A) Retrospective auditing B) Concurrent auditing
C) Prospective auditing D) Comprehensive auditing

Answer: C) Prospective auditing. Prospective auditing examines processes and controls before services are provided or claims are submitted, offering the best opportunity to prevent compliance violations.

203. What is the most important factor in audit sample selection? A) Convenience of record access B) Statistical validity and representativeness C) Availability of staff time D) Cost of conducting the audit

Answer: B) Statistical validity and representativeness. Audit samples should be statistically valid and representative of the population being examined to ensure audit results accurately reflect overall compliance performance.

204. What confidence level is commonly used in healthcare compliance audits? A) 85% B) 90% C) 95% D) 99%

Answer: C) 95%. Healthcare compliance audits commonly use 95% confidence levels, meaning auditors can be 95% confident that audit results reflect true population performance within specified error margins.

205. Which audit methodology provides real-time compliance assessment? A) Retrospective auditing B) Concurrent auditing C) Prospective auditing
D) Annual auditing

Answer: B) Concurrent auditing. Concurrent auditing occurs during service delivery, providing real-time compliance assessment and enabling immediate correction of identified problems.

206. What is the primary benefit of risk-based audit planning? A) Reduced audit costs B) Faster audit completion C) Focus on highest-risk areas for maximum impact D) Simplified audit procedures

Answer: C) Focus on highest-risk areas for maximum impact. Risk-based audit planning focuses limited audit resources on areas with highest probability and impact of compliance violations, maximizing audit effectiveness and value.

207. Which documentation is most important in compliance auditing? A) Audit planning documents B) Medical records supporting billed services C) Staff training certificates D) Policy acknowledgment forms

Answer: B) Medical records supporting billed services. Medical records that document and support billed services are typically the most important documentation in compliance auditing, providing evidence of service delivery and medical necessity.

208. What should be the first step in audit finding development? A) Determining corrective actions needed B) Calculating financial impact C) Gathering and analyzing evidence D) Reporting findings to management

Answer: C) Gathering and analyzing evidence. Audit finding development should begin with thorough evidence gathering and analysis to ensure findings are accurate, complete, and properly supported before developing conclusions.

209. What is root cause analysis in compliance auditing? A) Identifying who made errors B) Calculating financial impact of problems C) Examining underlying reasons why problems occurred D) Determining regulatory violations

Answer: C) Examining underlying reasons why problems occurred. Root cause analysis examines the underlying system problems and contributing factors that allowed compliance issues to occur, guiding effective corrective action development.

210. How should audit findings be prioritized? A) By dollar amount of overpayments identified B) By ease of correction C) By risk level and potential impact D) By regulatory agency interest

Answer: C) By risk level and potential impact. Audit findings should be prioritized based on their risk level and potential impact on patients, compliance, and organizational operations rather than just financial considerations.

211. What is the most effective approach for corrective action planning? A) Address symptoms of problems quickly B) Focus on individual accountability and discipline C) Address root causes and system improvements D) Implement minimum changes required

Answer: C) Address root causes and system improvements. Effective corrective action plans address root causes and implement system improvements that prevent problem recurrence rather than just addressing symptoms or individual errors.

212. How often should high-risk areas be audited? A) Monthly B) Quarterly C) Annually D) Based on risk assessment and previous findings

Answer: D) Based on risk assessment and previous findings. Audit frequency for high-risk areas should be determined by risk assessment results, previous audit findings, and regulatory requirements rather than arbitrary schedules.

213. What is the primary purpose of audit follow-up activities? A) Verifying corrective actions were implemented B) Finding additional problems C) Ensuring sustained compliance improvement D) Documenting audit completion

Answer: C) Ensuring sustained compliance improvement. Audit follow-up should verify that corrective actions were implemented effectively and have achieved sustained compliance improvement rather than just temporary fixes.

214. Which factor is most important for audit program credibility? A) Number of audits conducted B) Dollar amount of overpayments identified C) Objectivity and accuracy of audit results D) Speed of audit completion

Answer: C) Objectivity and accuracy of audit results. Audit program credibility depends primarily on the objectivity and accuracy of audit results rather than activity levels or financial recoveries identified.

215. What is the most common cause of audit program failure? A) Insufficient audit staff B) Lack of management support C) Inadequate audit technology D) Complex regulatory requirements

Answer: B) Lack of management support. Lack of management support is the most common cause of audit program failure, as effective programs require organizational commitment and resource allocation from leadership.

216. Which approach is most effective for communicating audit results? A) Written reports only B) Verbal presentations only C) Multiple communication methods with clear action items D) Email summaries with attachments

Answer: C) Multiple communication methods with clear action items. Effective audit communication uses multiple methods including written reports, presentations, and follow-up discussions with clear action items and accountability for implementation.

217. What is the purpose of compliance risk assessment? A) Meeting regulatory requirements B) Identifying and prioritizing areas for compliance focus C) Documenting organizational policies D) Preparing for external audits

Answer: B) Identifying and prioritizing areas for compliance focus. Compliance risk assessment identifies and prioritizes areas where compliance problems are most likely to occur or would have the greatest impact, guiding resource allocation and program focus.

218. Which framework is most commonly used for healthcare risk management? A) ISO 31000 B) COSO Enterprise Risk Management

C) NIST Risk Management Framework D) Joint Commission Risk Standards

Answer: B) COSO Enterprise Risk Management. COSO Enterprise Risk Management framework is widely used in healthcare for its integrated approach to risk management across organizational functions and strategic objectives.

219. What does COSO stand for in risk management? A) Committee of Sponsoring Organizations B) Compliance and Safety Oversight Standards C) Corporate Standards and Oversight Organization D) Clinical Operations and Safety Organization

Answer: A) Committee of Sponsoring Organizations. COSO stands for the Committee of Sponsoring Organizations of the Treadway Commission, which developed widely-used frameworks for internal control and enterprise risk management.

220. Which risk domain is unique to healthcare organizations? A) Financial risks B) Operational risks C) Clinical and patient safety risks D) Strategic risks

Answer: C) Clinical and patient safety risks. Clinical and patient safety risks are unique to healthcare organizations, requiring specialized knowledge and approaches that other industries typically don't face.

221. What is the purpose of risk matrices in compliance programs? A) Documenting all organizational risks B) Visualizing risk likelihood and impact for prioritization C) Meeting regulatory reporting requirements D) Training staff on risk identification

Answer: B) Visualizing risk likelihood and impact for prioritization. Risk matrices provide visual representations of risk likelihood and impact that help organizations prioritize risks and allocate resources to areas of greatest concern.

222. Which factor should be considered in addition to likelihood and impact when assessing risks? A) Historical precedent B) Risk velocity and detectability C) Regulatory interest D) Industry benchmarks

Answer: B) Risk velocity and detectability. Risk velocity (how quickly risks can escalate) and detectability (how easily risks can be identified) are important additional factors that affect risk prioritization and mitigation strategies.

223. What is the most effective type of risk control? A) Detective controls that identify problems B) Corrective controls that fix problems C) Preventive controls that stop problems before they occur D) Administrative controls that document problems

Answer: C) Preventive controls that stop problems before they occur. Preventive controls are most effective because they stop problems before they occur, eliminating potential harm and compliance violations rather than just detecting or correcting them.

224. Which approach is most effective for measuring compliance program effectiveness? A) Count of compliance activities completed B) Number of violations found and corrected C) Trend improvements in key risk indicators D) Regulatory feedback and inspection results

Answer: C) Trend improvements in key risk indicators. Measuring trend improvements in key risk indicators provides the best assessment of compliance program effectiveness by showing sustained improvement in compliance performance over time.

225. What is the primary purpose of compliance monitoring? A) Meeting regulatory requirements B) Ongoing surveillance for early problem detection C) Documenting compliance activities D) Preparing for external reviews

Answer: B) Ongoing surveillance for early problem detection. Compliance monitoring provides ongoing surveillance that enables early detection of potential problems before they become serious violations, supporting proactive compliance management.

226. Which monitoring approach provides the most comprehensive coverage? A) Transaction-based monitoring B) Risk-based monitoring focusing on high-risk areas C) Random sampling across all activities D) Complaint-driven monitoring

Answer: B) Risk-based monitoring focusing on high-risk areas. Risk-based monitoring provides the most effective coverage by focusing resources on areas with highest probability and impact of problems while maintaining adequate oversight of other areas.

227. What is the most important characteristic of effective risk mitigation strategies? A) Low implementation cost B) Quick implementation timeline C) Address root causes of risks D) Minimal operational disruption

Answer: C) Address root causes of risks. Effective risk mitigation strategies address root causes of risks rather than just symptoms, creating lasting risk reduction rather than temporary fixes that allow problems to recur.

228. How should organizations measure risk management performance? A) Number of risks identified B) Risk register completion C) Reduction in risk exposure and incident rates D) Risk assessment completion rates

Answer: C) Reduction in risk exposure and incident rates. Risk management performance should be measured by actual reduction in risk exposure and incident rates rather than activity measures like assessments completed or risks identified.

229. What is the relationship between compliance and risk management? A) They are completely separate functions B) Compliance is a subset of risk management C) Risk management is a subset of compliance D) They overlap significantly but have distinct focuses

Answer: D) They overlap significantly but have distinct focuses. Compliance and risk management overlap significantly in healthcare organizations but have distinct focuses - compliance on regulatory adherence and risk management on broader organizational threats and opportunities.

230. Which stakeholder group is most important for risk management program success? A) Board of directors B) Senior management C) Department managers D) All stakeholders are equally important

Answer: D) All stakeholders are equally important. Successful risk management requires engagement from all stakeholder groups, from board oversight to front-line staff involvement in risk identification and mitigation.

231. What is the most common weakness in healthcare compliance auditing? A) Inadequate sample sizes B) Lack of follow-up on corrective actions C) Insufficient documentation of findings D) Limited scope of audit coverage

Answer: B) Lack of follow-up on corrective actions. Many compliance audit programs fail to conduct adequate follow-up to ensure corrective actions are implemented effectively and achieve sustained compliance improvement.

232. Which audit approach is most cost-effective for routine compliance monitoring? A) Comprehensive annual audits B) Focused audits based on risk indicators C) Continuous monitoring with automated tools D) External audits by consulting firms

Answer: C) Continuous monitoring with automated tools. Continuous monitoring using automated tools provides cost-effective routine compliance surveillance while enabling focused audit attention on areas where problems are identified.

233. What is the primary benefit of using data analytics in compliance auditing? A) Reduced audit costs B) Faster audit completion C) Improved problem detection and pattern identification D) Simplified audit reporting

Answer: C) Improved problem detection and pattern identification. Data analytics enhance compliance auditing by improving problem detection and identifying patterns that might not be apparent through traditional auditing approaches.

234. Which documentation should be prioritized in compliance audit workpapers? A) Evidence supporting audit conclusions B) Detailed audit procedures performed C) Staff time spent on audit activities D) Comparison to previous audit results

Answer: A) Evidence supporting audit conclusions. Audit workpapers should prioritize documentation of evidence supporting audit conclusions, ensuring findings can be substantiated and defended if questioned.

235. What is the most effective way to ensure audit independence? A) Using external auditors only B) Separate reporting lines for audit staff C) Rotating audit staff assignments D) Independence policies and organizational structure

Answer: D) Independence policies and organizational structure. Audit independence requires both clear policies and organizational structures that prevent conflicts of interest and ensure auditors can reach objective conclusions.

236. Which factor is most important for audit quality assurance? A) Supervisor review of audit work B) Standardized audit procedures C) Regular training for audit staff D) All factors are equally important

Answer: D) All factors are equally important. Audit quality assurance requires supervisor review, standardized procedures, staff training, and other quality controls working together to ensure consistent, reliable audit results.

237. What should be the primary focus of compliance trend analysis? A) Comparing performance to other organizations B) Identifying patterns that indicate emerging risks C) Documenting historical compliance performance D) Meeting regulatory reporting requirements

Answer: B) Identifying patterns that indicate emerging risks. Compliance trend analysis should focus on identifying patterns and trends that might indicate emerging risks requiring proactive attention rather than just historical documentation.

238. Which approach is most effective for corrective action tracking? A) Annual reviews of action plan status B) Systematic tracking with regular status updates C) Informal follow-up with responsible parties D) External verification of completion

Answer: B) Systematic tracking with regular status updates. Effective corrective action tracking requires systematic approaches with regular status updates and accountability for implementation rather than periodic reviews or informal follow-up.

239. What is the most important outcome of effective risk assessment? A) Complete documentation of all risks B) Prioritized action plans for highest risks C) Regulatory compliance demonstration D) Stakeholder communication materials

Answer: B) Prioritized action plans for highest risks. Effective risk assessment should result in prioritized action plans that address the highest risks first, enabling focused resource allocation and systematic risk reduction.

240. Which risk management approach is most appropriate for healthcare organizations? A) Risk avoidance through conservative practices B) Risk transfer through insurance coverage C) Risk mitigation through systematic controls D) Integrated approach using multiple strategies

Answer: D) Integrated approach using multiple strategies. Healthcare organizations should use integrated risk management approaches that combine avoidance, mitigation, transfer, and acceptance strategies depending on specific risks and organizational circumstances.

241. What is the primary purpose of compliance key performance indicators? A) Demonstrating activity levels B) Measuring program effectiveness and improvement C) Satisfying regulatory requirements D) Benchmarking against other organizations

Answer: B) Measuring program effectiveness and improvement. Compliance KPIs should measure program effectiveness and improvement rather than just activity levels, providing meaningful information about compliance performance and outcomes.

242. Which factor most influences the success of compliance corrective actions? A) Speed of implementation B) Cost of corrective measures C) Leadership support and accountability D) Complexity of solutions implemented

Answer: C) Leadership support and accountability. Leadership support and accountability for corrective action implementation are the most important factors in success, providing necessary resources and organizational commitment.

243. What is the most effective approach for compliance program evaluation? A) Annual self-assessments B) External reviews by consultants C) Regular measurement against established objectives D) Regulatory inspection results

Answer: C) Regular measurement against established objectives. Effective compliance program evaluation requires regular measurement against established objectives and performance standards rather than periodic assessments or external validation alone.

244. Which compliance monitoring approach provides the earliest warning of problems? A) Annual compliance audits B) Real-time automated monitoring C) Complaint-based investigations D) Regulatory inspection findings

Answer: B) Real-time automated monitoring. Real-time automated monitoring provides the earliest warning of potential compliance problems by identifying issues as they occur rather than waiting for periodic reviews or external identification.

245. What should be the primary goal of compliance risk mitigation? A) Eliminating all compliance risks B) Reducing risks to acceptable levels C) Transferring risks to other parties D) Documenting risk mitigation efforts

Answer: B) Reducing risks to acceptable levels. Compliance risk mitigation should focus on reducing risks to acceptable levels rather than attempting to eliminate all risks, which is typically neither possible nor cost-effective.

246. Which approach is most effective for sustaining compliance improvements? A) Periodic reinforcement training B) Regular monitoring and feedback C) Updated policies and procedures D) Integration into organizational culture and processes

Answer: D) Integration into organizational culture and processes. Sustaining compliance improvements requires integration into organizational culture and daily processes rather than relying on periodic interventions or policy updates alone.

247. What is the most important factor in compliance program sustainability? A) Adequate funding and resources B) Regulatory stability C) Leadership commitment over time D) Staff expertise and training

Answer: C) Leadership commitment over time. Leadership commitment over time is the most important factor in compliance program sustainability, providing consistent support through changes in personnel, priorities, and external pressures.

248. Which measurement approach provides the best assessment of compliance program maturity? A) Comparison to regulatory minimum requirements B) Benchmarking against peer organizations C) Assessment against best practice frameworks D) Evaluation of program component completeness

Answer: C) Assessment against best practice frameworks. Assessment against established best practice frameworks provides the most meaningful evaluation of compliance program maturity and opportunities for advancement.

249. What should be the ultimate objective of healthcare compliance programs? A) Avoiding regulatory violations and penalties B) Protecting patients while supporting organizational mission C) Meeting minimum regulatory requirements D) Demonstrating good faith compliance efforts

Answer: B) Protecting patients while supporting organizational mission. The ultimate objective of healthcare compliance programs is protecting patients while supporting organizational mission, with regulatory compliance serving as a means to achieve these broader goals.

250. Which factor is most critical for long-term compliance program success? A) Advanced technology and automation B) Comprehensive policies and procedures C) Organizational culture that values ethical behavior D) Adequate staffing and resources

Answer: C) Organizational culture that values ethical behavior. Organizational culture that genuinely values ethical behavior and patient protection is the most critical factor for long-term compliance program success, enabling sustained compliance even as personnel, regulations, and circumstances change.

Professional Development and Emerging Areas

251. What is the most widely recognized healthcare compliance certification? A) CHC (Certified in Healthcare Compliance) B) CCEP (Certified Compliance and Ethics Professional) C) CCS (Certified Coding Specialist) D) CHPS (Certified in Healthcare Privacy and Security)

Answer: A) CHC (Certified in Healthcare Compliance). The CHC (Certified in Healthcare Compliance) from HCCA is the most widely recognized certification specifically focused on healthcare compliance, covering the broadest range of compliance topics.

252. How many years of relevant experience are typically required for CHC certification? A) 1 year B) 2 years C) 3 years D) 5 years

Answer: B) 2 years. CHC certification typically requires at least 2 years of relevant healthcare compliance experience, though some educational background may substitute for portions of the experience requirement.

253. How often must CHC certification be renewed? A) Annually B) Every 2 years C) Every 3 years D) Every 5 years

Answer: B) Every 2 years. CHC certification must be renewed every 2 years through continuing education requirements and payment of renewal fees to maintain the credential.

254. What organization offers the CCEP-HC certification? A) HCCA (Health Care Compliance Association) B) SCCE (Society of

Corporate Compliance and Ethics) C) AHIMA (American Health Information Management Association) D) AAPC (American Academy of Professional Coders)

Answer: B) SCCE (Society of Corporate Compliance and Ethics). The CCEP-HC (Certified Compliance and Ethics Professional - Healthcare) is offered by SCCE, focusing on compliance and ethics program development and management.

255. Which certification focuses specifically on research compliance? A) CHC (Certified in Healthcare Compliance) B) CHRC (Certified in Healthcare Research Compliance) C) CIP (Certified IRB Professional) D) Both B and C

Answer: D) Both B and C. Both CHRC (Certified in Healthcare Research Compliance) and CIP (Certified IRB Professional) focus on research compliance, with CHRC covering broader research compliance and CIP focusing specifically on IRB functions.

256. What is the primary focus of CHPS certification? A) General healthcare compliance B) Healthcare privacy and security C) Healthcare billing and coding D) Healthcare quality management

Answer: B) Healthcare privacy and security. CHPS (Certified in Healthcare Privacy and Security) from AHIMA focuses specifically on HIPAA privacy and security requirements and related healthcare data protection issues.

257. Which degree is most valuable for healthcare compliance career advancement? A) Master of Health Administration (MHA) B) Master of Business Administration (MBA) C) Juris Doctor (JD) D) All provide valuable but different benefits

Answer: D) All provide valuable but different benefits. MHA, MBA, and JD degrees all provide valuable benefits for healthcare compliance careers, with MHA focusing on healthcare management, MBA on business skills, and JD on legal analysis.

258. What is the average salary increase associated with professional certification? A) 5-10% B) 10-15% C) 15-25% D) 25-35%

Answer: C) 15-25%. Professional certification in healthcare compliance is typically associated with 15-25% salary increases, though individual results vary based on experience, location, and organizational factors.

259. Which professional organization has the largest membership in healthcare compliance? A) HCCA (Health Care Compliance Association) B) SCCE (Society of Corporate Compliance and Ethics) C) AHLA (American Health Lawyers Association) D) AHIMA (American Health Information Management Association)

Answer: A) HCCA (Health Care Compliance Association). HCCA has the largest membership specifically focused on healthcare compliance, though SCCE covers broader compliance topics across multiple industries.

260. What is the most effective study strategy for certification examinations? A) Intensive study in the weeks before the exam B) Systematic study over several months with practice tests C) Group study sessions with peers D) Professional review courses only

Answer: B) Systematic study over several months with practice tests. Systematic study over several months combined with regular practice testing provides the most effective preparation for certification examinations by allowing adequate time for knowledge development and retention.

261. Which continuing education activities provide the most career value? A) Online webinars and courses B) Professional conferences and networking C) Academic coursework and degrees D) All activities provide different but valuable benefits

Answer: D) All activities provide different but valuable benefits. Different continuing education activities provide various benefits - webinars for current topics, conferences for networking, and academic coursework for deep knowledge - requiring balanced approaches for optimal career development.

262. What is telehealth's primary compliance challenge? A) Technology security requirements B) Multi-state licensing and regulatory requirements C) Patient privacy protection D) Billing and reimbursement accuracy

Answer: B) Multi-state licensing and regulatory requirements. Multi-state licensing and varying regulatory requirements across jurisdictions create the most complex compliance challenges for telehealth services, affecting provider licensure and practice standards.

263. Which federal agency has primary oversight for telehealth prescribing regulations? A) FDA (Food and Drug Administration) B) DEA (Drug Enforcement Administration) C) CMS (Centers for Medicare & Medicaid Services) D) OIG (Office of Inspector General)

Answer: B) DEA (Drug Enforcement Administration). The DEA has primary oversight for controlled substance prescribing regulations that significantly impact telehealth services, particularly through the Ryan Haight Act requirements.

264. What does the Ryan Haight Act primarily regulate? A) Telehealth technology standards B) Controlled substance prescribing via telemedicine C) Telehealth reimbursement rates D) Interstate medical practice

Answer: B) Controlled substance prescribing via telemedicine. The Ryan Haight Act specifically regulates controlled substance prescribing via telemedicine, generally requiring in-person medical evaluations before prescribing controlled substances.

265. Which requirement is most challenging for telehealth HIPAA compliance? A) Patient consent procedures B) Technology platform security C) Business associate agreements with platform vendors D) Documentation requirements

Answer: C) Business associate agreements with platform vendors. Business associate agreements with telehealth platform vendors can be challenging because many platforms were not originally designed for healthcare use and may not provide adequate HIPAA compliance features.

266. What is information blocking under federal regulations? A) Preventing unauthorized access to health information B) Practices that interfere with health information access, exchange, or use C) Cybersecurity measures for protecting patient data D) Privacy protections for sensitive health information

Answer: B) Practices that interfere with health information access, exchange, or use. Information blocking refers to practices that are likely to interfere with the access, exchange, or use of electronic health information, prohibited under federal regulations promoting interoperability.

267. How many exceptions exist to information blocking prohibitions? A) 6 exceptions B) 7 exceptions C) 8 exceptions D) 10 exceptions

Answer: C) 8 exceptions. Information blocking regulations include 8 exceptions that allow certain practices when they serve legitimate purposes such as privacy, security, safety, performance, content and manner, fees, licensing, and infeasibility.

268. Which actors are subject to information blocking prohibitions? A) Healthcare providers only B) Health information exchanges only C) Healthcare providers, HIEs, and health IT developers D) All healthcare organizations

Answer: C) Healthcare providers, HIEs, and health IT developers. Information blocking prohibitions apply to healthcare providers using certified health IT, health information exchanges, and health IT developers and networks.

269. What is required for patient access under information blocking rules? A) Patients must request access in writing B) Electronic access without special effort C) Access only through healthcare provider portals D) Reasonable fees may be charged for all access

Answer: B) Electronic access without special effort. Information blocking rules require that patients be given access to their electronic health information without special effort, typically through patient portals or similar electronic access methods.

270. Which penalty can be imposed for information blocking violations? A) Criminal prosecution only B) Civil monetary penalties up to $1 million per violation C) Exclusion from federal programs D) Loss of health IT certification

Answer: B) Civil monetary penalties up to $1 million per violation. Information blocking violations can result in civil monetary penalties up to $1 million per violation, making compliance a significant financial concern for covered actors.

271. What does ESG stand for in healthcare compliance? A) Electronic Security Guidelines B) Environmental, Social, and Governance C) Emergency Safety Guidelines D) Ethical Standards and Governance

Answer: B) Environmental, Social, and Governance. ESG stands for Environmental, Social, and Governance factors that increasingly affect healthcare organizations through stakeholder expectations, regulatory requirements, and access to capital.

272. Which environmental issue is most relevant to healthcare compliance? A) Carbon emissions from facilities B) Medical waste disposal and management C) Water conservation programs D) Energy efficiency improvements

Answer: B) Medical waste disposal and management. Medical waste disposal and management represents the most direct environmental compliance issue for healthcare organizations, with specific regulatory requirements for handling and disposing of potentially infectious materials.

273. What social responsibility requirement applies to nonprofit hospitals? A) Charity care provision B) Community health needs assessment C) Health equity reporting D) All of the above

Answer: D) All of the above. Nonprofit hospitals face multiple social responsibility requirements including charity care provision, community health needs assessments, and increasing expectations for health equity initiatives and reporting.

274. Which governance factor is most important for healthcare ESG compliance? A) Board diversity and independence B) Executive compensation alignment C) Transparency and stakeholder engagement D) All factors are equally important

Answer: D) All factors are equally important. Healthcare ESG governance requires attention to board composition, executive compensation, transparency, and stakeholder engagement as interconnected elements of responsible organizational governance.

275. What is the primary driver of increased ESG focus in healthcare? A) New federal regulations B) Stakeholder and investor expectations C) Accreditation requirements D) Patient demand for transparency

Answer: B) Stakeholder and investor expectations. Increased ESG focus in healthcare is primarily driven by stakeholder and investor expectations for responsible organizational behavior, though regulatory requirements are also developing.

276. Which technology trend most affects healthcare compliance? A) Artificial intelligence and machine learning B) Cloud computing and data storage C) Mobile health applications D) Robotic process automation

Answer: A) Artificial intelligence and machine learning. Artificial intelligence and machine learning have the most significant impact on healthcare compliance by changing how organizations process information, make decisions, and manage risks.

277. What is the primary compliance concern with AI in healthcare? A) Data privacy and security B) Bias and discrimination in algorithms C) Regulatory approval requirements D) All concerns are equally important

Answer: D) All concerns are equally important. AI in healthcare creates multiple compliance concerns including data protection, algorithmic bias, regulatory requirements, and clinical decision support standards that must be addressed comprehensively.

278. Which regulatory framework is most relevant to healthcare AI governance? A) FDA medical device regulations B) HIPAA privacy and security requirements C) Anti-discrimination laws D) All frameworks apply to different aspects

Answer: D) All frameworks apply to different aspects. Healthcare AI governance must consider FDA device regulations for clinical AI, HIPAA for data protection, anti-discrimination laws for algorithmic fairness, and other relevant regulatory frameworks.

279. What is the most important factor for successful telehealth compliance programs? A) Technology platform selection B) Multi-state regulatory analysis and compliance C) Staff training on telehealth procedures D) Patient education and consent processes

Answer: B) Multi-state regulatory analysis and compliance. Multi-state regulatory analysis and compliance represents the most complex and important factor for telehealth compliance, affecting provider licensing, practice standards, and legal liability.

280. Which career path offers the best opportunities for healthcare compliance professionals? A) Traditional hospital-based compliance roles B) Consulting and advisory services C) Health technology and digital health companies D) All paths offer good opportunities in different ways

Answer: D) All paths offer good opportunities in different ways. Different career paths offer various opportunities - hospitals provide foundational experience, consulting offers diverse exposure, and health technology companies offer innovation opportunities requiring different skills and interests.

281. What is the fastest-growing area of healthcare compliance? A) Privacy and cybersecurity B) Value-based care and quality reporting C) Digital health and health technology D) Traditional billing and coding compliance

Answer: C) Digital health and health technology. Digital health and health technology represent the fastest-growing areas of healthcare compliance as organizations adopt new technologies that create novel compliance challenges and regulatory requirements.

282. Which skill is most important for future healthcare compliance professionals? A) Technical regulatory knowledge B) Data analysis and technology skills C) Communication and relationship building D) All skills remain important with evolving emphasis

Answer: D) All skills remain important with evolving emphasis. Future healthcare compliance professionals need technical knowledge, data skills, and communication abilities, with increasing emphasis on technology and data analysis capabilities alongside traditional compliance expertise.

283. What is the most significant trend affecting healthcare compliance education? A) Increased online and virtual learning options B) Integration of technology and data analytics training C) Focus on practical application over theoretical knowledge D) All trends are reshaping compliance education

Answer: D) All trends are reshaping compliance education. Healthcare compliance education is being reshaped by online learning expansion, technology integration, and practical application focus, requiring educational programs to adapt to changing professional needs.

284. Which professional development activity provides the best return on investment? A) Professional certification B) Advanced degree programs C) Industry networking and relationships D) ROI varies based on individual circumstances and goals

Answer: D) ROI varies based on individual circumstances and goals. Professional development ROI varies significantly based on individual career goals, current position, geographic location, and personal circumstances rather than having universal best options.

285. What is the most important factor for compliance career longevity? A) Technical expertise in current regulations B) Continuous learning and adaptation to change C) Strong relationships with regulatory agencies D) Specialization in specific compliance areas

Answer: B) Continuous learning and adaptation to change. Continuous learning and adaptation to change is most important for compliance career longevity as regulations, technology, and healthcare delivery continue evolving rapidly.

286. Which compliance area requires the most frequent updates to knowledge and skills? A) HIPAA privacy and security B) Fraud and abuse laws C) Health information technology and cybersecurity D) Quality reporting and value-based care

Answer: C) Health information technology and cybersecurity. Health information technology and cybersecurity require the most frequent knowledge updates due to rapidly evolving technology, emerging threats, and changing regulatory responses.

287. What is the most effective approach for staying current with regulatory changes? A) Reading regulatory publications regularly B) Attending professional conferences annually C) Multiple information sources with systematic monitoring D) Relying on organizational legal counsel updates

Answer: C) Multiple information sources with systematic monitoring. Staying current with regulatory changes requires multiple information sources including regulatory publications, professional organizations, conferences, and peer networks with systematic monitoring approaches.

288. Which factor most influences compensation in healthcare compliance careers? A) Years of experience B) Geographic location C) Organization size and complexity D) All factors significantly influence compensation

Answer: D) All factors significantly influence compensation. Healthcare compliance compensation is influenced by experience, location, organization size, education, certification, and specific role responsibilities rather than any single dominant factor.

289. What is the most common career progression path in healthcare compliance? A) Analyst to manager to director to chief compliance officer B) Specialist to generalist to leadership roles C) Multiple paths exist with various progression possibilities D) Linear progression through increasing responsibility levels

Answer: C) Multiple paths exist with various progression possibilities. Healthcare compliance offers multiple career progression paths including specialist tracks, generalist advancement, cross-functional moves, and consulting opportunities rather than single linear progressions.

290. Which geographic region offers the best opportunities for healthcare compliance professionals? A) Major metropolitan areas with large health systems B) Regions with significant pharmaceutical industry presence C) Areas with academic medical centers D) Opportunities exist in all regions with different characteristics

Answer: D) Opportunities exist in all regions with different characteristics. Healthcare compliance opportunities exist throughout the country with different characteristics - urban areas offer more positions, rural areas offer broader responsibilities, and various regions have industry concentrations.

291. What is the most important consideration when changing healthcare compliance positions? A) Salary and compensation increases B) Career advancement opportunities C) Alignment with career goals and organizational culture D) Geographic location and family considerations

Answer: C) Alignment with career goals and organizational culture. Career moves should prioritize alignment with long-term career goals and organizational culture fit over short-term financial gains or other single factors for sustainable career satisfaction.

292. Which networking approach is most effective for healthcare compliance professionals? A) Professional organization participation B) Industry conference attendance C) Online professional networks and social media D) Comprehensive approach using multiple networking methods

Answer: D) Comprehensive approach using multiple networking methods. Effective networking requires comprehensive approaches using professional organizations, conferences, online networks, mentorship relationships, and other methods to build diverse professional relationships.

293. What is the most valuable benefit of professional mentorship in healthcare compliance? A) Access to job opportunities B) Technical knowledge transfer C) Career guidance and perspective D) Professional reference and recommendations

Answer: C) Career guidance and perspective. Professional mentorship provides valuable career guidance and perspective that helps professionals navigate complex career decisions and understand industry dynamics beyond technical knowledge.

294. Which factor is most important for work-life balance in healthcare compliance careers? A) Organization culture and expectations B) Role responsibilities and complexity C) Industry sector and setting D) Personal boundary setting and time management

Answer: A) Organization culture and expectations. Organization culture and leadership expectations most significantly influence work-life balance, as some organizations support reasonable schedules while others create unsustainable demands regardless of role or industry.

295. What is the most significant challenge facing healthcare compliance professionals? A) Keeping up with regulatory changes B) Managing increasing complexity and workload C) Balancing compliance requirements with operational needs D) All challenges are significant and interconnected

Answer: D) All challenges are significant and interconnected. Healthcare compliance professionals face interconnected challenges including regulatory complexity, workload management, and operational balance that require comprehensive approaches rather than focusing on single issues.

296. Which trend will most impact the future of healthcare compliance? A) Artificial intelligence and automation B) Value-based care and quality focus C) Digital health and remote care delivery D) All trends will significantly impact the field

Answer: D) All trends will significantly impact the field. The future of healthcare compliance will be shaped by AI and automation, value-based care evolution, digital health expansion, and other interconnected trends requiring adaptable professional approaches.

297. What is the most important advice for new healthcare compliance professionals? A) Focus on technical regulatory knowledge B) Build relationships throughout the organization C) Develop both technical skills and business understanding D) Seek broad experience across multiple compliance areas

Answer: C) Develop both technical skills and business understanding. New healthcare compliance professionals should develop both technical regulatory knowledge and business understanding to be effective partners who can translate compliance requirements into practical organizational solutions.

298. Which personal characteristic is most important for healthcare compliance success? A) Attention to detail and accuracy B) Communication and interpersonal skills C) Ethical behavior and professional integrity D) All characteristics are essential for comprehensive success

Answer: D) All characteristics are essential for comprehensive success. Healthcare compliance success requires attention to detail, communication skills, ethical behavior, and other characteristics working together rather than any single dominant trait.

299. What is the most rewarding aspect of healthcare compliance careers? A) Financial compensation and benefits B) Job security and career stability C) Protecting patients and supporting healthcare mission D) Professional recognition and advancement opportunities

Answer: C) Protecting patients and supporting healthcare mission. Most healthcare compliance professionals find protecting patients and supporting healthcare mission to be the most rewarding aspect of their careers, providing meaning beyond financial or professional benefits.

300. What is the key to long-term success in healthcare compliance? A) Technical expertise and regulatory knowledge B) Professional networking and relationship building C) Continuous learning and ethical commitment D) Adaptability and resilience in changing environments

Answer: C) Continuous learning and ethical commitment. Long-term success in healthcare compliance requires continuous learning to stay current with changes and unwavering ethical commitment that maintains professional credibility and personal satisfaction throughout career challenges and opportunities.

References

The following sources were referenced and cited throughout this healthcare compliance certification guide:

Professional Organizations and Certifying Bodies

- American Academy of Professional Coders (AAPC)
- American College of Physicians (ACP)
- American Health Information Management Association (AHIMA)
- American Hospital Association (AHIA)
- American Medical Association (AMA)
- American Society for Healthcare Risk Management (ASHRM)
- Association of Corporate Counsel (ACC)
- Health Care Compliance Association (HCCA)
- Society of Corporate Compliance and Ethics (SCCE)

Government Agencies and Official Sources

- Centers for Medicare & Medicaid Services (CMS)
- Food and Drug Administration (FDA)
- U.S. Department of Health and Human Services (HHS)
- U.S. Department of Justice (DOJ)
- Office for Civil Rights (OCR)
- Office of Inspector General (OIG)

Academic and Educational Institutions

- Drexel Online
- Kent State University
- Northeastern University
- Seton Hall Law School
- University of Pittsburgh Law School
- USC Gould School of Law

Healthcare Organizations and Case Studies

- Anthem Inc.
- Banner Health
- Cleveland Clinic
- Geisinger Health System
- Halifax Health System
- Intermountain Healthcare
- Kaiser Permanente
- Mayo Clinic
- Mercy Health System
- Methodist Health System
- Novartis Pharmaceutical
- Orlando Health
- Premera Blue Cross
- Regional Health Network
- Riverside Regional Medical Center
- Tenet Healthcare Corporation
- Tuomey Healthcare System
- Tufts Medical Center
- University of Rochester Medical Center
- West Boca Medical Center

Legal and Compliance Firms

- Akerman LLP
- Day Pitney
- Holt Law
- Morgan Lewis
- Ogletree Deakin
- Venable LLP
- Womble Bond Dickinson

Technology and Service Providers

- AuditBoard

- Carepatron
- ComplianceJunction
- Compliancy Group
- CyberArrow
- MDaudit
- MedSafe
- MedTrainer
- Narola Infotech
- NAVEX
- Symplr

Publications and Research Organizations

- ADEC Healthcare
- American Medical Compliance
- Ethico
- HIPAA Journal
- Institute of Internal Auditors (IIA)
- MRO Corp
- Net Health

International and Regulatory Bodies

- European Data Protection Supervisor
- GDPR Register
- GDPR Local

Specialized Resources

- Committee of Sponsoring Organizations (COSO)
- Healthcare Financial Management Association (HFMA)
- Joint Commission
- LegalNodes
- MDinteractive
- Medisolv
- Mometrix Test Preparation

This reference list encompasses the diverse sources consulted in developing this healthcare compliance certification guide, representing authoritative voices from government agencies, professional organizations, academic institutions, healthcare systems, legal firms, and industry experts who contribute to the evolving field of healthcare compliance.

Made in United States
North Haven, CT
04 November 2026

21736089R00143